MATH GEEK

From Klein Bottles to Chaos Theory,

MATH

a Guide to the Nerdiest Math Facts,

GEEK

Theorems, and Equations

RAPHAEL ROSEN

Avon, Massachusetts

Published by
Adams Media, a division of F+W Media, Inc.
57 Littlefield Street, Avon, MA 02322. U.S.A.
www.adamsmedia.com

ISBN 10: 1-4405-8381-1
ISBN 13: 978-1-4405-8381-0
eISBN 10: 1-4405-8382-X
eISBN 13: 978-1-4405-8382-7

Printed in the United States of America.

10   9   8   7   6   5   4   3   2   1

Library of Congress Cataloging-in-Publication Data

Rosen, Raphael
  Math geek / Raphael Rosen.
    pages cm
  Includes index.
  ISBN 978-1-4405-8381-0 (pb) -- ISBN 1-4405-8381-1 (pb) -- ISBN 978-1-4405-8382-7 (ebook) -- ISBN 1-4405-8382-X (ebook)
  1. Mathematics. I. Title.
  QA39.3.R675 2015
  510--dc23
                            2015005078

Many of the designations used by manufacturers and sellers to distinguish their products are claimed as trademarks. Where those designations appear in this book and F+W Media, Inc. was aware of a trademark claim, the designations have been printed with initial capital letters.

Romanesco broccoli photo © iStockphoto.com/Picture Partners; Pine cone photo © iStockphoto.com/ tamara_kulikova; spirograph image © iStockphoto.com/Kaprinay; tessellation image © iStockphoto .com/AlexanderZam; platonic solids and Möbius strip images © iStockphoto.com/john woodcock; sextant image © iStockphoto.com/nicoolay.

Dimples on golf ball image based on work by Prakul varshney (own work) via Wikimedia Commons; golf ball image © iStockphoto.com/Blankstock.

Golden ratio image by Ahecht (Original); Pbroks13 (Derivative work); Joo. (Editing) [CC0 1.0 (http:// creativecommons.org/publicdomain/zero/1.0/deed.en), via Wikimedia Commons.

Raindrop images by Pbroks13 (own work) [CC BY 3.0 (http://creativecommons.org/licenses/by/3.0/ deed.en), via Wikimedia Commons.

Origami fold images by Heron2, via Wikimedia Commons.

Sudoku grid from *The Everything® Puzzles for Commuters Book* by Charles Timmerman, copyright © 2007 by F+W Media, Inc., ISBN 10: 1-59869-406-5, ISBN 13: 978-1-59869-406-2.

Cover images © iStockphoto.com/pashabo/VLADGRIN/blueringmedia; Alexey Buhantsov/123RF.

DEDICATION

To Nathaniel, Jolina, and the rest of my family

CONTENTS

Contents

PART 2: BEHAVIOR   89

PART 3: PATTERNS

Contents

PART 4: SPECIAL NUMBERS   211

ACKNOWLEDGMENTS

I wouldn't have been able to write this book without the help of many people. I especially want to thank Dave Auckly, professor of mathematics at Kansas State University, and Francis Su, the Benediktsson-Karwa Professor of Mathematics at Harvey Mudd College, for their time and help. When I was lost in a mathematical thicket, their clear explanations showed me the way out. And I of course want to thank my editors, who supported me throughout the writing process.

I also want to thank Jolina and Nathaniel for being patient while I devoted long hours to completing this project. I love you always.

INTRODUCTION

WHAT DOES IT MEAN TO BE A MATH GEEK? Maybe you enjoyed math classes when you were in school and now do logic puzzles in your spare time. Maybe you've become intrigued by all of the references to math in popular culture—*Proof, Numb3rs, The Imitation Game, A Beautiful Mind*—and want to know more. Maybe you're an engineer or a physicist and use advanced math concepts every day. Maybe you have a hard time understanding math and yearn to get a glimpse of a world that so many people find fascinating. Or maybe you're your own kind of geek: after all, there are as many kinds of math geeks as there are theorems.

Whoever you are, in these pages I hope to show you that mathematics is not just a series of rote exercises performed in a classroom. You won't have to memorize anything, and there's no test at the end. I hope to convince you that mathematics is something built into the fabric of reality: a collection of shapes, patterns, numbers, arguments, and, well, little treasures. Math is in the air you breathe, the sidewalks you walk on, and the buses you take to work each morning. What does that mean? To find out, you'll have to read on.

Besides showing you that mathematics is a living feature of the world we live in, I also hope to persuade you that math is pretty. I'm not saying that equations look good on paper, or that plus and minus signs are like calligraphy. I mean that learning about math is like looking at a sunset, reading a poem, or listening to your favorite band. Math has a beauty that can stop you in your tracks. Have you ever walked out of a movie theater after seeing a great drama, your mind stunned by the performances, sets, and cinematography? Believe it or not, math is like that. Some mathematicians have even argued that math should be included in a list of cultural touchstones that includes Shakespeare, Mozart, and Michelangelo. These math mavens believe that people should study math for its own sake, because *not* studying it would be a crime on a par with never having read *Hamlet*. In other words, people should not learn about mathematics simply to get a good score on the SAT. Instead, they should study it to enrich their lives.

Our expedition to find math in our everyday world will take us from pizza to donuts, from online shopping to the GPS features in our smartphones. We'll look closely at why, when you're waiting at a bus stop, there won't be a bus for what seems like an eternity, and then suddenly two or three will arrive all at once. We'll stop to examine weird vegetables in your local supermarket, and learn how music is translated into a file on your iPod. We'll even make sense of strange paradoxes like why adding roads can make traffic worse.

Once you learn the fascinating math concepts hiding in the world around you, you'll find an even greater appreciation for math—one that you can share with your fellow commuter when the bus is late . . . again.

PART 1
SHAPES

ROMANESCO BROCCOLI

The Beauty of Romanesco Broccoli

MATHEMATICAL CONCEPT: SELF-SIMILARITY

HAVE YOU EVER looked closely at the fruits and vegetables in your local supermarket? Some of them are creepy: the yellow Buddha's hand, for instance, looks like a squid creature from an H.P. Lovecraft story. Others are strangely beautiful. Sweet potatoes have a wonderful lumpiness like misshapen hunks of clay; onions have nesting rings like those found in trees; and when you slice open an apple, you can see that the seeds are arranged in a star-shaped configuration that's oddly pleasing. Even ornamental cabbage—which is sold at garden stores—has a kind of geometric attractiveness.

But nothing in the produce aisle beats the Romanesco broccoli for the ultimate in vegetable beauty. In fact, it's hard to take your eyes off it. The Romanesco, a type of *Brassica oleracea*, or cabbage, has the general shape of a pinecone, but its surface is a riot of smaller pinecone shapes, and on the surface of each of those cones are *more* cones, and so on. Each of the smaller surface cones looks like the larger, original cone, so much so that if you were to lop off a surface cone, photograph it, and place the photo next to an image of the entire broccoli, it would be difficult to tell which was which.

Mathematicians would say that the shape of the Romanesco is *self-similar:* if you zoom in on the shape and look closely at a detail, what you see is the same as if you hadn't zoomed in at all. Self-similarity—when an object looks the same no matter the scale—is also a distinguishing feature of fractals, a kind of shape studied and popularized by mathematician Benoit Mandelbrot. His 1982 book, *The Fractal Geometry of Nature*, helped introduce this species of object to the world. (The book was largely a revision of his 1977 book called *Fractals: Form, Chance and Dimension*.) Mandelbrot identified lots of shapes in nature that have this self-similar quality: jagged coastlines, clouds, and the exquisite tracery of veins in a leaf. Nature seems to like self-similar shapes; the more you look for them, the more you find.

THE MANDELBROT SET

Benoit Mandelbrot also studied something that's now called the Mandelbrot set, a set of complex numbers in a sequence that does not approach infinity. When you plot the Mandelbrot set on a graph, it can take on a bulbous shape with a captivating beauty, which is interesting to mathematicians in part because the more you zoom in on any part of it, the more detail you see. In fact, as you zoom, you start to see the original Mandelbrot set shape over and over again.

2

Measuring the Length of a Coastline: Not as Easy as It Sounds

MATHEMATICAL CONCEPT: MEASUREMENT

WHAT COULD BE MORE straightforward than measuring how long something is? If we want to figure out the length of a table, for instance, we can use a measuring tape. If we want to determine the distance from one town to another, we can drive a car and note the change in the odometer. Or we can get a road map, use a ruler to measure the distance between the two towns, and then use the map's scale to convert inches to miles, or centimeters to kilometers.

But measuring the length of a coastline is more complicated. It turns out that the length of any particular coastline depends on the length of the unit used to measure it. In general, the smaller the measuring unit, the longer the coastline. And in principle, as the measuring unit gets smaller and smaller, the length of a coastline increases to infinity. How can this be?

Like many forms in nature, coastlines are irregular and jagged. Thus, as you zoom in on one you notice more and more detail. If you were peering down at North America from the height of a satellite, for example, the coastline would appear relatively smooth and feature-free. But if you were measuring a coastline by walking, you would observe river inlets, small spits of land, and rocks, among other features. Then, if you were to get on your hands and knees, you could start to account for pebbles and leaves. And if you were to use a microscope, your measurements could include molecules. At each new level of detail, the unit of measurement shrinks, from kilometer to meter to centimeter to micrometer; and each time, the amount of territory to measure increases. If you were to measure the coastline of Great Britain using a stick 100 kilometers (about 62 miles) long, the resulting distance would be more like 2,800 kilometers (approximately 1,700 miles). But if you reduced your measuring stick to 50 kilometers (31 miles), the new total coastline distance would be 3,400 kilometers (2,100 miles).

The coastline paradox illustrates how even though mathematics makes possible measurements of extraordinary precision, it can also reveal fuzziness inherent in the very structure of reality.

CANADIAN COAST

Canada has the longest coastline of any country in the world, at 152,100 miles. But just imagine how much longer it might be if you were to measure the whole thing using a yardstick.

3

Bubbles Are Fun
and Efficient

MATHEMATICAL CONCEPT: VOLUME

IMAGINE A SUNNY DAY in the park during the summer. Chances are, a kid there is playing with soap bubbles. Whether you make them using a plastic wand or a large hoop made out of straws and string, soap bubbles—with their shimmery surfaces and blobby shapes—are the airborne embodiment of fun.

They are also a wellspring of mathematical pondering. Mathematicians have known for a long time that if you want to enclose a given volume of air in a shape with the least surface area, that shape is a sphere. But what if you want to enclose *two* volumes of air? The hunch was that the best way was to use a double bubble. A double bubble is the shape that's formed when two bubbles join. (You've probably seen one if you've ever taken a bubble bath.) Usually, the bubbles are separated by a flat membrane; if one bubble is larger than the other one, then the membrane pushes slightly into the bigger bubble. In 1995, mathematicians Joel Hass, Michael Hutchings, and Roger Schlafly published a paper with a proof that the double-bubble shape is the most efficient way to enclose two *equal* volumes of air. But what if the two volumes of air aren't equal? Is the double-bubble shape still the way to enclose them using the least amount of surface area?

The answer is yes. In 2000, mathematicians Frank Morgan, Michael Hutchings, Manuel Ritoré, and Antonio Ros published a proof that gave a much more general result, showing that the double-bubble shape is the best way to enclose *any* two volumes of air using the least amount of surface area. They showed that the double bubble uses less surface area than countless other configurations that two joined bubbles can take, including a weird case in which one bubble wraps around the other's midsection, like a donut. (In mathematics, the donut shape has a special name—torus—and pops up in the subfield of topology.) Moreover, this math team created their proof without using a computer.

This is one of those cases when mathematics can use reason to probe the workings of nature to learn its secrets. All you need is paper and a pencil.

THE MARANGONI EFFECT

Soap bubbles last longer than bubbles made of other materials, like pure water, because of the Marangoni effect, which describes the flow of material along the boundaries of areas with different degrees of surface tension. It's named for Italian physicist Carlo Marangoni, who published his findings in 1865. Basically, when it comes to soap, the Marangoni effect actually stabilizes the bubble's boundaries, making it stronger and longer lasting than a regular bubble.

Is There Math Behind Jackson Pollock's Paintings?

MATHEMATICAL CONCEPT: FRACTALS

JACKSON POLLOCK created some of the most iconic paintings of the twentieth century, and some researchers have argued that their appeal stems from mathematics. Specifically, scientists have argued that the drip paintings Pollock completed in the 1940s incorporate fractals, which are geometric patterns that repeat at both small and large scales. Some also argue that Pollock's works are especially captivating because they seem to capture some of the fractal qualities of our environment. (Fractals occur frequently in nature, like within the texture of a cloud.)

Fractals have dimensions, just as lines (one dimension) and beach balls (three dimensions) do, but unlike those objects, fractals often have dimensions that include decimals. In general, mathematicians categorize fractal dimensions according to a scale that goes from 0 to 3. Some one-dimensional fractals, like a segmented line, have fractal dimensions of 0.1 to 0.9. Two-dimensional fractals, like the contour of a coastline, have fractal dimensions ranging from 1.1 to 1.9. And three-dimensional fractals, like a head of cauliflower, have fractal dimensions of 2.1 to 2.9.

In the late 1990s, physicist Richard Taylor noticed that Pollock's drip paintings seemed to have fractal properties, and proposed that one could measure the fractal characteristics of Pollock's work. Using a specific kind of analysis, a person could conceivably figure out whether any painting was created by Pollock. Taylor's technique involved scanning photographs of Pollock's paintings into a computer and then overlaying a grid on top of the digital images. The computer then analyzed the painting, comparing patterns in boxes as large as the entire painting and as small as a fraction of an inch. Taylor found that Pollock's painting did indeed incorporate fractals. For instance, one painting—*Number 14*—was determined to have a fractal dimension of 1.45, which matches the dimension of many coastlines.

Years later, though, researchers at Case Western Reserve University in Cleveland found evidence that Taylor's technique did not seem to reliably identify Pollock's work. One doctoral student discovered that a rough sketch of stars she created using Adobe Photoshop passed Taylor's test. Another study found that two paintings by Case Western undergraduates also passed Taylor's test, while two authentic Pollocks did not. The researchers concluded that the test did not include a wide enough range of boxes to sufficiently determine whether any particular painting was a Pollock.

PIET MONDRIAN

For a more straightforward example of math in art, check out the paintings of Piet Mondrian, who used straight lines and quadrilaterals to great effect in his work.

5

The Koch Snowflake

MATHEMATICAL CONCEPT: FRACTALS

THERE IS AN ODDNESS about fractals (see entry #4), a strangeness that is hard to explain but easy to show with examples. One example is the Koch snowflake, a shape based on the Koch curve, which was first described by Swedish mathematician Niels Fabian Helge von Koch. To create a Koch snowflake, begin with an equilateral triangle (one whose sides all have the same length). Now, divide each side into three equal parts. Using the middle segment of each side, form another equilateral triangle whose base is the middle segment and which points outward. Continue this process forever.

One strange result of this process is that the Koch snowflake ends up having an infinite length. Each time you construct a new triangle in the middle of one of the snowflake's sides, or legs, you increase the length by one-third. And since the process goes on forever, so does the growth of the snowflake's perimeter.

Here's another strange result: Though the perimeter increases boundlessly, getting larger and larger without end, the area enclosed by the snowflake, though always increasing, *does* have a boundary. If you think of a circle drawn around the original triangle, you can see that the area of the Koch snowflake will never grow larger than the area of the circle. It may approach the circle's area, but will never surpass it. So in a sense, a mathematical object with infinite length encloses a finite area. Weird!

THE CESÀRO FRACTAL

Some fractals are formed not by addition, but instead by subtraction. You create the Koch snowflake by adding peaks to the centers of line segments, but to create a variation known as the Cesàro fractal, you *take away* peaks. The result is a snowflake that looks like it's been chewed by a shark. In the end, though, the more complex they both become, the more similar they look to the human eye.

6

Are You Living in the Fourth Dimension?

MATHEMATICAL CONCEPTS: KLEIN BOTTLES, GEOMETRY, TOPOLOGY

KLEIN BOTTLES ARE WEIRD. Well, let me explain a little more thoroughly. Understanding them requires envisioning the fourth dimension—a kind of space that exists at right angles to our three-dimensional space—but though they are strange, Klein bottles might hold the secret to the fate of our universe.

First described by German mathematician Felix Klein in 1882, the Klein bottle's original German name was *Kleinsche Fläche*, or Klein surface, but may have been misunderstood as *Kleinsche Flasche*, or Klein bottle. In any case, the name stuck. A Klein bottle is actually a surface—a 2D manifold—and like a sphere, a Klein bottle has no boundary. It also is non-orientable, meaning that directions change as you move along the surface.

But Klein bottles are famous for another reason: they have no inside or outside. The two actually merge into one space. (You can think of Klein bottles as being analogous to Möbius strips—see entry #7—which have only one side. In fact, if you were to slice a

Klein bottle in two, you would end up with two Möbius strips.) The other notable fact about Klein bottles is that they cannot exist in three-dimensional space. To create one out of, say, a sheet of paper, you would first roll the paper into a cylinder. Then instead of joining the two ends together, forming a donut, you would give one end a twist. That twist cannot be performed unless you "lift" one end of the cylinder into the fourth dimension. Because we live in a three-dimensional world, the best we can do is push one end of the cylinder through the body of the cylinder and join the twisted end to the other end. The resulting shape intersects itself, but if we were four-dimensional beings, the Klein bottle wouldn't intersect at all.

To see why, imagine that you live in two-dimensional space. Now imagine there is a bounded line in that space, like a two-dimensional rope. If someone asked you to form that line into the shape of a figure eight, but avoid having the line intersect itself, you'd have no idea what to do. How would that be possible? To do so, you'd have to somehow "lift" the line into three-dimensional space; at that point, the shape could be formed with no intersections.

Back to the relationship between Klein bottles and the fate of the universe. The future of the universe—including the destinies of stars, galaxies, and even space itself—depends in part on the universe's overall shape. Scientists have suggested many possible shapes that are consistent with their observations, and while some of those shapes resemble a flat sheet of paper extending in all directions forever—a three-dimensional space known as a Euclidean 3-manifold—others are "closed," meaning that while they are very large, they eventually turn in on themselves. (One example of a closed shape is a sphere. If you begin at any point on the surface of a sphere and travel along the surface in a straight line, you will eventually return to your starting point.) For all we know, though, the universe could be shaped very differently. Just as we live on a spherical object but our immediate environment suggests to our senses that we live on an infinitely large flat plane, our local place in the universe suggests that the universe extends in straight lines in all directions, but in truth, at distances too far to observe, the universe might look like a saddle or cylinder. Or it could be shaped like a Klein bottle.

So if you thought that the fourth dimension had nothing to do with your everyday life, think again. In fact, you might be living in it.

FELIX KLEIN

Born in 1849, Felix Klein taught mathematics at Germany's University of Göttingen and had deep interests in geometry. He was also famous for marrying the granddaughter of philosopher Georg Wilhelm Friedrich Hegel!

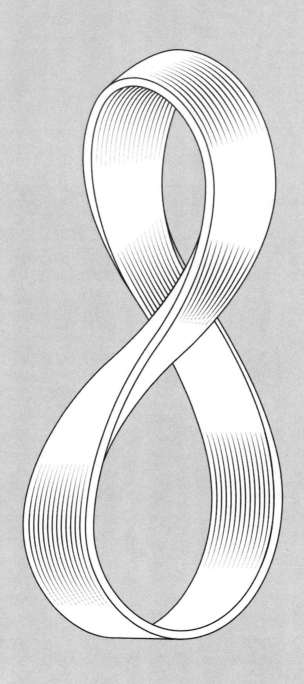

7

Building a Better Conveyor Belt

MATHEMATICAL CONCEPTS: MÖBIUS STRIP, TOPOLOGY

IN MATHEMATICS, little things can have big consequences. For example, take a strip of paper, of any length. Hold an end in each hand and twist the paper 180 degrees. Now glue the ends of the strip together. You have just constructed a true mathematical oddity, using nothing but basic office supplies. The object you've created is known as a Möbius strip.

Möbius strips are special because, in mathematics, they are non-orientable, meaning that they have only one side. While that may sound impossible, you can prove the one-sidedness to yourself. Take a pencil and start drawing a line at any part of the strip and just keep going. (Make sure you draw the line parallel to the sides of the strip so the pencil doesn't run off the paper into the air.) Eventually the pencil will end up at its starting point. And, significantly, it will have covered the entire surface of the strip. If the strip had had two sides—an outside and an inside—the pencil line would only have marked one or the other, leaving the remaining side untouched.

This weird, one-sided object sounds exotic—and it is—but Möbius strips turn up now and again in the world outside math books and chalkboards. For example, in 1957 the B.F. Goodrich Company created a Möbius-strip conveyor belt. The half-twist ensured that both sides of the belt would be used, saving wear and tear. The same reasoning is behind why some recording tapes and typewriter ribbons were also designed to have a Möbius shape: more surface area could be used, increasing the usability of the product. Möbius strips also appear in the world of electronics—specifically, in the shape of some resistors (which provide resistance in an electronic circuit)—and in biology: some configurations of molecules have a Möbius-strip structure.

The Möbius strip was named after August Ferdinand Möbius, the nineteenth-century German mathematician who discovered it. (It turns out that the strip was discovered at almost the same time by Johann Benedict Listing, another nineteenth-century German mathematician who coined the mathematical term "topology.") Möbius had a distinctive lineage: his ancestors include Martin Luther, one of the religious thinkers who helped start the Protestant Reformation in the early sixteenth century, and he studied with Carl Friedrich Gauss, one of the most accomplished mathematicians in history.

The Möbius strip is a great example of a simple object that anyone can create but that also has deep mathematical implications. And there's nothing like the feeling you get while holding math in your hands.

MUSICAL DYADS

Music and mathematics have an interesting connection. Music theorists sometimes graph on a sheet of paper how various musical two-note chords—thirds, fourths, fifths, octaves—are related, while taking into account that you can write a dyad two ways (C-F or F-C, for example). Capturing this relationship on a piece of paper requires twisting and turning it into a Möbius strip.

8

The Mathematical Connection Between Your Shoelaces and Your DNA

MATHEMATICAL CONCEPTS: KNOT THEORY, CURVES

YOU WOULDN'T EXPECT to find mathematics in a pair of shoes. But take a moment to glance down at your knotted shoelaces. Those twisted loops can actually be a gateway to complex mathematical thought.

The branch of mathematics at issue here is known as knot theory. Knots in math, though, are different from knots in your everyday experience in one significant way: they don't have any loose ends. (In math language, these knots are closed.) You can, in fact, make a closed knot for yourself. Take a piece of string—or a piece of wet linguini or a lasso—and tie a standard square knot. Now, take the two loose ends and join them with a piece of tape. What you end up with might look like a soft pretzel, but a knot it is!

Though parts of it are familiar, knot theory has its peculiarities. The definition of a knot in mathematics, as Colin Adams puts it in *The Knot Book*, is "a closed curve in space that does not intersect itself anywhere." That definition might lead you to wonder what knot is the *simplest* one you could make. That knot is a circle, and it goes by the fantastically cool, *Alice-in-Wonderland*-like name of an "unknot." (This knot is also sometimes known as the trivial knot.) After the unknot, the next simplest knots are the figure eight and trefoil.

What exactly happens in a day in the life of a knot theorist? They tend to be interested in things like whether one could untangle a given knot without cutting it, or whether by messing around with a knot one could reveal that it's actually the unknot in an unfamiliar form. But knot theory affects more people than mathematicians. Biologists are interested in knot theory because DNA, the molecule that encodes the materials that make up living creatures, can sometimes be knotted, and those knots can affect how the DNA molecule's information is interpreted by an organism's cellular machinery. Chemists, too, are interested in knots. Many of them would like to tinker with knotted molecules, because the kind of a knot a particular molecule might have could completely alter how that molecule behaves. (One configuration might mean that the substance behaves like an oil; another might mean that the substance acts like a gel.) Just a twist or two could have a dramatic impact.

TAIT CONJECTURES

Nineteenth-century mathematician Peter Guthrie Tait categorized knots according to how many crossings they had. He also came up with three conjectures that involve alternating (crossing under and over to form the link), chiral knots (those that are *not* equivalent to their mirror images), and the writhe of a knot (a geometric quantity that describes coiling in the knots). All three of his conjectures have recently been proven to be true.

What the Subway
Map Leaves Out

MATHEMATICAL CONCEPT: TOPOLOGY

EXAMINE ALMOST ANY SUBWAY MAP in any city in the world. What do you observe? Unlike maps in an atlas, which show a road's every twist and turn, subway maps are relatively simple. They consist of straight lines, circles, and gentle curves. (Take a look at the maps of London's Tube, Boston's T, or Washington, D.C.'s Metro to see some examples.) Subways rarely travel along such uncomplicated routes, though: the trains traverse a series of wiggles and waggles while moving from one station to the next. But despite this discrepancy, subway maps still help travelers navigate. How is it that these maps can leave out so much information and still be useful?

The answer lies within a field of mathematics known as topology. Related to geometry, topology deals with how shapes change when they are stretched, shrunken, pulled, twisted, and distorted. (The word "topology" comes from Greek words meaning place and study or measurement. The changes studied by topology must follow a rule, though: the changes can't violate the original shape's integrity. For instance, shapes that are cut and glued back together are not considered appropriate subjects for topological study. On the other hand, the new shapes created when you pull a rubber band to its limit, scrunch it into a ball, and twist it into a pretzel are all fair game. In short, in topology you must be able to turn a new shape back into its original shape in one continuous motion. If you can, then as far as topology is concerned, the two shapes are equivalent.

Now the relationship between a subway map and the subway's actual route becomes clear. A subway map is a topological transformation of the physical subway path. In a sense, the map shows a version of the subway route that has been stretched and smoothed out, as if it were made of Silly Putty. According to topology, the two shapes—the map route and the route as it actually exists in the public transportation system—are the same.

THE LARGEST SUBWAY(S) IN THE WORLD

The Shanghai Metro in China is the longest subway system in the world by route length—its tracks cover more than 330 miles. But the New York City Subway has the most stops of any in the world, with an official count of 468 stations.

Origami

MATHEMATICAL CONCEPTS: GEOMETRY, TOPOLOGY

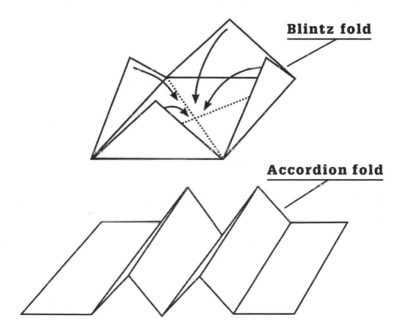

Blintz fold

Accordion fold

ORIGAMI, the Japanese art of folding paper, has a reputation in the United States as a pastime for children. Many of us have seen origami paper cranes, drinking cups, and air-filled balloons. What is less well known is that origami has a strong connection to mathematics.

One fascinating aspect of origami is its power to go beyond traditional mathematics, especially geometry. Using nothing but creased paper, a person can trisect an angle (divide it into three equal subangles), a task that is impossible using the compass and straightedge of traditional geometry. A person can also use origami to double a cube, another task that standard geometry cannot accomplish. (Cube-doubling is a problem that was known to the ancient Egyptians and Greeks. To double a cube, one must begin with a cube with a given side length and volume, and construct a new cube that has twice the volume of the original. This procedure cannot be completed because the side of the larger cube would have a length of the cube-root of 2, a length that cannot be constructed using a compass and straightedge.)

In fact, the mathematical study of origami has yielded its own geometric axioms, a set of principles and definitions similar to those devised by Euclid, the renowned mathematician who lived in Greece more than 2,000 years ago. These seven tenets are known as the Huzita-Hatori axioms; they list all the ways one can perform a single fold. Origami mathematics has also resulted in Kawasaki's theorem, which maintains that in a set of angles surrounding a single point, the sum of every other angle equals 180 degrees.

Besides nearly becoming its own mathematical realm, complete with its own proofs and axioms, the *subject matter* of origami is often mathematical. Some people make three-dimensional forms out of modular origami components that are shaped like triangles and pentagons. Some people make origami versions of the Platonic solids, the five basic shapes composed of regular polyhedra (three-dimensional shapes made up of flat faces with straight sides). Other origamists construct hyperbolic parabolas, saddle-shaped forms that look like a cross between a square and a butterfly. Finally, some people use origami to prove the Pythagorean theorem.

Origami and mathematics seem to, in a sense, share the same conceptual DNA. And there's nothing like building a form with your own hands to get a better grasp on a mathematical concept. Forget pencils and graphing calculators; find time to discover math in folding paper!

ORIGAMI HOLIDAY TREE

Each year, in partnership with OrigamiUSA, the American Museum of Natural History produces the Origami Holiday Tree. The tree has approximately 800 origami ornaments. In 2014 the theme was based on the Night at the Museum movies, so the ornaments included Theodore Roosevelt, a *Tyrannosaurus rex*, and a statue from Easter Island.

11

There's Math Behind Your Tangled Cords

MATHEMATICAL CONCEPT: KNOT THEORY

IT'S ONE OF THE ANNOYANCES of modern life. You reach into your pocket or bag for your iPhone earbuds, only to find that they have tangled themselves into an impossible-to-untie knot. You retrieve your garden hose from its resting place in the basement to find that, lo and behold, it has worked itself into a knot. You unpack your Christmas lights from a box in the attic to discover that, what else, they have twisted themselves into a ball of knots. Why do so many objects in our lives tend to end up in knots, despite our efforts to prevent their doing so?

It turns out that there is a mathematical reason that long, flexible things like cords and strings tend to become knotted. In fact, two physicists at the University of California, San Diego, published a research paper in 2007 on this very topic. In essence, there are only a few possible configurations in which a bunched-up string-like object remains knot-free—for instance, when the length of the string remains parallel to itself at all times, at no point touching itself or crossing over itself—but many, many configurations in which the string starts to become knotted. Actually, it can take only a few seconds for a string or cord to get tangled. All that has to happen is that one of the loose ends crosses over one part of the cord's length. At that point it's easy for the loose end to braid itself with the rest of the string.

During their research, the UC San Diego team tumbled pieces of string of various lengths for ten seconds inside a rotating box attached to a motor. They analyzed the resulting knots using mathematical knot theory, trying to find the mathematical equation (in this case, the Jones polynomial) that corresponded to each kind of knot. (Knot theory classifies knots according to how many crossings they have.) They found that around 96% of the knots that resulted from the tumbling were "prime knots," or knots with a minimum number of crossings ranging from 3 to 11. The team also found that knots formed less often when the string was shorter—under half a meter—but when its length approached 2 to 6 meters, the probability of the string becoming knotted grew rapidly, up to 50%. Beyond those lengths, the probability did not increase significantly.

So though you may curse your earbuds the next time you have to painstakingly untangle them from a balled-up mess, try to appreciate the math behind it.

ANTI-TANGLE DEVICES

Tangled phone cords spawned an entire industry. In the days when people relied on telephone receivers that were tethered to their bases, inventors created anti-tangle devices ranging from a part that can swivel 360 degrees to a tube that slides into the spaces inside the cord's curls, all meant to prevent this everyday irritation.

12

Why Bicycle Gears Are Different Sizes

MATHEMATICAL CONCEPTS: GEOMETRY, RATIOS

ONCE UPON A TIME, bicycles looked goofy. In the nineteenth century, bicycles had enormous front wheels and tiny rear wheels. The pedals were attached directly to the front wheel, which could be almost five feet in diameter, and the rider would have to jump onto the seat as if he were mounting a horse. These kinds of bikes soon went out of fashion, in part because if the bike hit a bump, the rider could easily be launched over the handlebars. Later, manufacturers began building bikes with gears and chains, a development that not only allowed the rider to sit in the middle of the bike, improving balance, but also made it possible to change gears, depending on the terrain. You might not need to change gears when riding your bike on flat ground, but when climbing a hill, changing to a different gear might mean the difference between riding the bike and pushing it. But how exactly do gears work? How do they make riding a bike uphill easier, or downhill more efficient?

The answer depends on ratios. When you connect a large gear to a small gear, turning one means that the other will turn, too, but at a different rate than the first. Let's say that the front gear on your bike is three times as large as the rear one. For every one rotation that the front gear makes, the rear gear will have to turn three times. You can think of it in terms of each wheel's circumference. (If you remember from math class, a circle's circumference equals pi times the circle's diameter.) If the front gear has a diameter of three inches, then its circumference is pi times 3, or around 9.42 inches. So if you were to draw a dot at one point of the gear's edge and then turn the gear one time, the dot would trace out a path in space that, if transferred to a piece of paper, would measure 9.42 inches long.

Now let's say that the rear gear measures one inch in diameter. Its circumference would measure 3.14 inches, and with every turn a dot on its edge would trace out a 3.14-inch path. But each time the front gear makes one revolution—9.42 inches—the rear gear has to turn three times. (Based on this difference in diameters, by the way, the gear ratio for these two gears is 3:1.)

You can therefore make the rear wheel rotate three times for every one rotation of the pedals (though you have to push three times as hard): perfect for speeding down hills.

GEAR TOYS

Gears aren't just useful; they're also fun to play with. Many toys on the market today feature gears, including Gears!Gears!Gears!, Gear & Rotor Fun, and the BlueLotus Rotatable Building Gears Sets. Some gear toys can be quite elaborate: Brickowl.com lists fifty-seven different kinds of gears for the LEGO Technic sets, including a gear with forty teeth, a gear with twenty-four teeth and an internal clutch, and a double-bevel gear with twenty teeth.

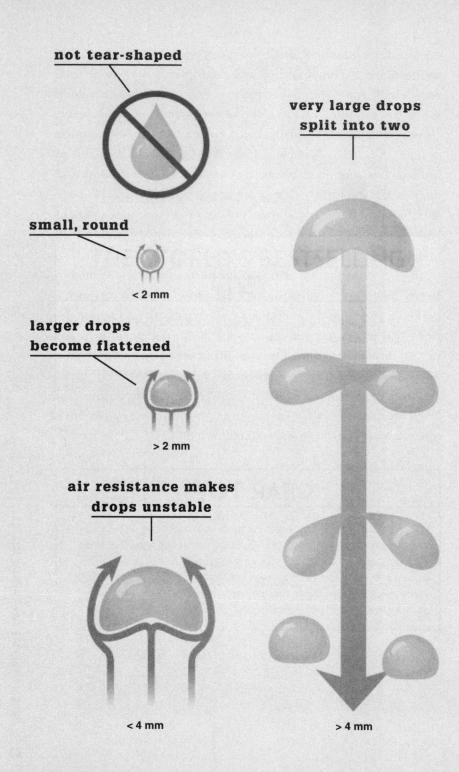

Myth Busted: Raindrops Aren't Shaped Like Teardrops

MATHEMATICAL CONCEPT: GEOMETRY

RAINDROPS are not what you think they are. At least, the shape of a raindrop is not what you've probably been led to believe. In cartoons, on weather maps, and in illustrations, the raindrop is usually shown looking like a classic tear, with a rounded bottom, and sides that taper upward to a point.

In reality, raindrops have a completely different shape. All raindrops begin as roughly spherical objects, as water in the atmosphere gloms on to particles of smoke and dust. Once the droplet has acquired enough heft, it starts falling. As it falls, the drop's surface tension—caused by hydrogen bonds between water molecules—holds the drop in a round shape. As the drop picks up speed, however, the pressure of the air pushing into the bottom of the drop makes the lower portion of the drop flat, like the bottom of a pan. At this point the raindrop looks more like the top half of a hamburger bun. If the drop gets too large, as it sometimes will when colliding with other drops on its journey to the earth, it will break apart into smaller drops—the breaking point seems to be around 4 millimeters in diameter.

THE CIRCUMFERENCE OF RAINDROPS

Raindrops vary in size. On average, a small drop during a light storm might have a circumference of about 0.5 millimeters, but during a heavy storm, a drop could be as big as 5 millimeters in circumference.

14

Why Are Traffic Signs Different Shapes?

MATHEMATICAL CONCEPT: SHAPES

EVERYONE KNOWS that stop signs are octagons; that is, they have eight equal sides. But not everyone knows *why* stop signs have this particular shape. Why eight sides? Why not three, or ten?

There are two reasons why traffic planners chose an eight-sided shape.

1. Unlike square signs, which used to be much more ubiquitous, an octagonal sign could be understood from multiple directions. Driver A would be able to know that driver B had a stop sign and was required to stop, even if driver A was approaching from another direction where the face of the sign was not visible.

2. Traffic engineers realized early on that not only the words printed on traffic signs, but the shapes of the signs themselves, could convey a message. So, they created standardized signs based on the idea that the more sides a sign's shape had, the more danger the sign indicated. For instance,

a round sign—which can be understood as having an unlimited number of sides—is used for train crossings. An octagon is used for intersections with other cars, and a four-sided diamond is used for warnings, to alert you to things like an upcoming S-shaped curve or a deer crossing.

So, the next time you're driving, pay attention to the geometry of the traffic signs all around you. Those shapes may save your life!

THE HISTORY OF STOP SIGNS

The first stop sign was installed in Detroit in 1915, and was a square sheet of metal with black lettering on a white background. But it was the Mississippi Valley Association of State Highway Departments that in 1923 made influential recommendations to add more sides to the stop sign's shape. And in 1935, the *Manual on Uniform Traffic Control Devices* recommended that stop signs be uniformly colored red.

15

Why Is the Pentagon Building Shaped Like That?

MATHEMATICAL CONCEPT: GEOMETRY

PENTAGONS ARE SHAPES, but when we talk about *the* Pentagon, we usually mean the building outside of Washington, D.C. that houses America's Defense Department. It happens to be one of the largest office buildings in the world, with twice the square footage of the Empire State Building. It houses approximately 25,000 employees. In fact, the U.S. Capitol building could fit in just one of the Pentagon's five sides. But why does the Pentagon have the shape it does?

As World War II was developing, the United States decided it needed a new facility to contain its burgeoning War Department. The site chosen was Arlington Farms, an experimental farm operated by the Department of Agriculture and located next to Arlington Cemetery, the resting place of active-duty soldiers and veterans. Because of the roads and other properties bordering the farm, the site had a roughly five-sided shape, so the original plans for the new War Department building accordingly fit that space. But officials soon grew uncomfortable with the idea of placing a military building so close to such a sensitive area and decided to move it to another site, which once had been the location of Hoover Field, the first airport to service Washington. It was too late to change the architectural plans, so while architects did tweak some design elements, the pentagonal shape remained.

Fortunately for all involved, the five-sided plans had advantages. Walking between any two points takes less than ten minutes, and architects could more easily distribute utilities throughout the building.

THE PENTAGON

The Pentagon covers approximately 583 acres, and has more than 6 million square feet of floor area. The building has seven floors . . . that we know of.

16

Triangles

MATHEMATICAL CONCEPTS: SHAPES, GEOMETRY

HAVE YOU EVER NOTICED how often triangles appear in your daily life? Whether you're biking to work (there's a triangle in the center of the bike frame) or cruising along an interstate highway (there are triangles in those enormous electricity transmission towers), triangles pop up again and again. Is there a reason, or do bike builders and civil engineers use these three-sided shapes on a whim?

It turns out that there is a very good reason why triangles feature prominently in so many aspects of our built environment. Triangles are uncommonly stable shapes, making them ideal choices for structures that must remain strong. Imagine a triangle whose three angles are hinges: even though the angles aren't rigid, the triangle will be. You could even imagine a triangle made up of bendy straws, with the bendy parts forming its angles. Despite the bendiness, the triangle will remain as stable as if its corners were made of solid plastic. Can you think of any other places where you have seen triangles?

TRIANGLE CONCERTO

The triangle—the musical instrument, that is—was first given a solo part in a piece of music in Franz Liszt's First Piano Concerto. One skeptical critic dubbed the piece the "Triangle Concerto."

17

Why Are Manhole Covers Round?

MATHEMATICAL CONCEPTS: SHAPES, GEOMETRY

THE SKY IS BLUE. Stones are hard. Grass is green. There are aspects of the world that we encounter every day, parts of our experience that are so common we hardly think about them at all. Sometimes mathematics can make us look at these everyday things in a different way, with new understanding and new insights.

One of those things is the manhole cover. Usually they are round, but why? Wouldn't any other shape do just as well?

It turns out that the circle is an ideal shape to cover a manhole because a circle is one of the only shapes that can't fall through a hole shaped like itself. To see why, imagine a manhole cover shaped like a triangle. And let's say that the triangle's legs aren't all the same length. (Maybe one is one foot long and the other two are two feet long each.) If we were to lift the manhole cover and rotate it upward so it was perpendicular to the surface of the street, the side with the shorter leg could fit through the hole. That's because two sides of the hole would be two feet long—to fit the triangle's two long legs—and an object one foot wide can fit through an opening that's two feet wide. The only way to prevent the cover from falling through its hole is to create a cover shaped in such a way that no matter how you turn it, no side is smaller or larger than any other side. A circle fits the bill perfectly.

MANHOLES IN SPACE

According to an urban legend, an underground 1950s nuclear test accidentally shot a manhole cover into space. The legend has never been corroborated, but it may stem from a true incident that occurred during Operation Plumbbob, a series of nuclear tests conducted throughout 1957. During one underground test, a 2,000-pound piece of metal was propelled into the air at an estimated speed of 66 kilometers per second (41 miles per second). It was never recovered.

18

LEGO Sets

MATHEMATICAL CONCEPT: COMPLEXITY

THE WORLD OF TOYS is another great place to find math. When I was a child, I thought LEGO sets were the best toys in the world. There was something profoundly pleasant about sitting in front of a tub of loose LEGO bricks and thinking about what to create next. It turns out, though, that LEGO sets are not just for play. They can also reveal aspects of mathematics that might not have been apparent otherwise.

Specifically, LEGO pieces play a role in the study of complexity. Mark Changizi and other researchers at the Duke Institute for Brain Sciences recently conducted a study to figure out the answer to a question that sounds deceptively simple. All systems have components: bodies have cells; computers have switches and processors; ecosystems have birds and trees. The researchers wanted to know if, when any system—whether it consists of animals, cells, or electronic parts—becomes larger, the number of *types* of components also increases. Compare the inner workings of a wristwatch to those of a grandfather clock. The clock will definitely have more individual pieces, but will those pieces fall into more categories than the watch pieces do?

The researchers confirmed that the number of categories in various kinds of systems *did* increase as the scale of an object increased. In fact, they graphed the results and found that the number of categories and the number of components had an interesting relationship. Specifically, the number of component types increased in proportion to the number of pieces, according to a power law. (A power law relationship looks like this: $Y = kX_a$. Y and X are the two variables we want to examine; in this case, Y stands for the number of components and X for the number of component types. In this formula, k is any number, called a constant, and a is an exponent of the X variable.) The team's research also showed that although the number of component types increased in conjunction with the number of pieces, the rate slowed as the number of pieces got higher and higher.

But when they counted the number of components and component types in LEGO sets, they discovered that the number of component types grew *faster* than those of other networks. In other words, the larger the LEGO set, the more types of pieces it has. LEGO enthusiasts, including Changizi himself, fear that this increase stems from a change in how LEGO designs its sets. These days, LEGO sets come in a large number of themes, including Star Wars, Ninjago, Teenage Mutant Ninja Turtles, and The Lord of the Rings, among others. This proliferation, some fear, has led to more LEGO-brick specialization, meaning that more pieces are best suited for only one theme, or even one set. For lovers of original LEGO parts, this change is a cause for sadness. But LEGO toys continue to be versatile in another sense: they appeal to a wide segment of the population, including mathematicians.

MASTER BUILDERS

Who designs the LEGO sets that you can buy in stores? LEGO Master Builders, that's who. These LEGO whizzes also design the life-sized models on view at LEGOLAND parks around the world. To become a Master Builder requires years of practice, and these experts are often first discovered through public competitions.

19

Let's Go Fly a . . . Quadrilateral

MATHEMATICAL CONCEPT: SHAPES

WITHOUT KITES, spring and summer just wouldn't be themselves. But their allure extends beyond the thrill of piloting a piece of fabric in a perfect breeze. Traditional American kites are a good example of a particular kind of quadrilateral (a shape with four sides). The basic kite has four sides, like a square or rectangle. But unlike those two shapes, a kite's sides are bunched together by length. So the two short sides are adjacent to each other, as are the two long sides. The placement of the sides is what gives the kite its distinctive elongated diamond shape.

The kite shape is also interesting because of the way in which many kites can fit together to cover a plane (which is just an idealized flat surface, like a piece of paper with no thickness). You can take any kind of kite, with any degree of angle between its sides, and use it, along with an infinite number of identical kites, to completely cover a plane, leaving no gaps between individual shapes. This kind of covering is called tiling. (Think of the tiles on your bathroom wall or floor and you'll have an idea of what's going on.)

Kites also play a role in Penrose tiling, a special kind of tiling in which the individual shapes form patterns that don't repeat in a regular way.

The shape of a kite also determines its ideal flying conditions. Diamond-shaped kites do best in light winds, but need tails to help them stabilize. Delta kites, which look like triangles, can fly in almost no wind. And six-sided rokkaku kites, which originated in Japan hundreds of years ago, are very maneuverable and are often used in kite battles. (If you cut the line of another kite or force it to the ground, you win!)

THE AREA OF A KITE

There are two ways to calculate the area of a kite. If you know the length of the two diagonals, then you can multiply the lengths and divide by 2. Or, if you know the length of one short and one long side, as well as the angle between them, you can use trigonometry, multiplying the short length by the long length, and then multiplying that product by the sine of the angle.

20

What Do Herpes and Table Salt Have in Common?

MATHEMATICAL CONCEPT: PLATONIC SOLIDS

NOT ALL THREE-DIMENSIONAL SHAPES are created equal. Think of any shape that exists, or could exist. Some, like the shape of a potato, are lumpy and irregular. Others, like a star, are neat, with tidy straight lines. Spheres are round and smooth, while the blocky shapes in *Tetris* have sharp edges.

Some shapes, though, are special. They have properties that have been studied for thousands of years. That historic group includes the Platonic solids. Named for the philosopher who lived in Athens in the fourth century B.C.E., these three-dimensional forms are constructed using two-dimensional shapes, like squares, triangles, and pentagons. But the two-dimensional shapes that make up a Platonic solid have to meet certain conditions.

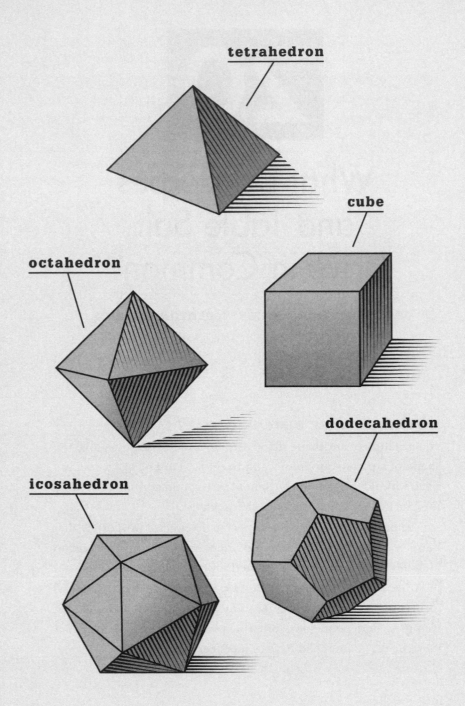

tetrahedron

cube

octahedron

dodecahedron

icosahedron

1. One, they have to be regular, meaning that all their lines have to be the same length and all their angles have to be the same size.
2. Two, they have to be congruent, meaning that all have to be identical, so that if you placed one on top of another they would match exactly. (In other words, you can't create a Platonic solid using triangles of different sizes.)
3. Third, at each vertex—the place on each shape where lines join up—the same number of shapes have to meet.

There are five, and only five, Platonic solids.
1. The tetrahedron has four faces, each one a triangle.
2. The cube is made up of six squares.
3. The octahedron has eight faces and is structured like two pyramids with their bases touching. (Like the tetrahedron, each face of an octahedron is a triangle.)
4. The dodecahedron has twelve faces, and each face is a pentagon.
5. The icosahedron has twenty faces, each one a triangle.

And if you're wondering why there are only five Platonic solids, Euclid—the ancient Greek mathematician—has an answer for you. He worked out a proof and included it in Book 13 of his *Elements*. Look it up if you're curious.

But these shapes weren't just considered to be mathematical curiosities. In Plato's dialogue *Timaeus*, the Greek philosopher claims (through one of his characters) that each solid corresponds to one of the elements that make up the natural world. The tetrahedron was associated with fire, the cube with earth, the octahedron with air, the icosahedron with water, and the dodecahedron with the arrangement of the constellations in the heavens.

Centuries later, in the late 1500s, Johannes Kepler used the Platonic solids to explain the structure of the solar system. Interested with figuring out why the planets were spaced out the way they were, Kepler posited that the distances could be modeled by placing a Platonic solid between the orbit of each planet, which was signified by a sphere. Beginning from the interior of the solar system, the order of the Platonic solids started with the octahedron, corresponding to the orbit of Mercury, then transitioned to the icosahedron, the dodecahedron, the tetrahedron, and the cube. (For Kepler, there were only five planets.)

While Kepler's explanation turned out to be incorrect, one insight was right on: Platonic solids are in fact a part of nature. For example:

- Many mineral crystals take the shape of cubes, including table salt (sodium chloride); if you were to walk along the shore of the Dead Sea, you would step on large salt cubes that had washed up from the sea's depths.
- Diamonds and fluorite often form crystals in the shape of octahedrons.
- Viruses, such as herpes, are often shaped like icosahedrons.
- Atoms often form bonds in the shape of a tetrahedron. Molecules of methane and ammonium ions are composed of four hydrogen atoms in a tetrahedral shape surrounding a carbon or nitrogen atom.

Platonic solids aren't just something in an ancient Greek tome—they are literally in the air we breathe and the ground on which we walk.

DIHEDRAL ANGLES

Each Platonic solid contains something called a dihedral angle, which is the interior angle between any two face planes. Because each face plane of a Platonic solid is the same, so are all the solid's dihedral angles. For example, in a cube, the dihedral angle is 90 degrees, just like the vertex angle. But in a tetrahedron, the dihedral angle is 70.6 degrees, while the vertex angle is 60. The larger the dihedral angle, the more closely the solid resembles a sphere.

21

Why Do Golf Balls Have Dimples?

MATHEMATICAL CONCEPTS: PHYSICS, GEOMETRY

WHEN YOU WATCH TIGER WOODS tee off during the U.S. Open, you might not imagine that mathematics is behind the scenes, helping his golf ball fly through the air. But it's true, and all thanks to the geometry of the ball's dimples.

Centuries ago, golf balls were wooden or rubber and had surfaces that were completely smooth. Golf myth maintains that as golfers used balls over and over again, they noticed that the older, dented balls flew farther than the newer, smooth balls did. Scientists later learned that the dents, or dimples, caused the air flowing over the ball to stay closer to the ball's curved shape, reducing the turbulence in the air behind the ball, which causes drag. Dimples have traditionally had the shape of circles, but some recent golf balls have dimples shaped like hexagons. Callaway, the manufacturer, claims that the hexagonal dimples cover more of the golf ball's surface, so there is less flat space between each dimple and thus even less drag.

Smooth Ball

Air quickly separates from ball.

Air flow around ball is laminar—layered and smooth.

A vortex is created. Swirling air creates heavy drag.

Golf Ball

Turbulence sucks air to ball. Separation is delayed.

Dimples create turbulence in layer of air around it.

This results in a smaller vortex and less drag.

DIMPLES

Golf balls vary in size, but most have between 300 and 500 dimples. A common golf ball has 336 dimples.

22

Gauss and Pizza

MATHEMATICAL CONCEPT: SHAPES

TRY THIS EXPERIMENT: take a sheet of newspaper and wrap a watermelon, as if you were planning on giving it to a friend as a birthday present. What do you find? No matter how hard you try, there will always be folds and creases that stick up; the paper never lies completely flush with the melon's surface. (To make the paper conform to the melon, you would have to take a pair of scissors and make a series of cuts, and even then you would probably have to smash down a stray fold here and there.) In fact, it's impossible to make a flat surface, like a sheet of paper, into a globe shape without bending or cutting or smashing.

The converse is just as difficult. Peel a grapefruit so you're left with one globe-shaped piece and try flattening it. The peel will inevitably break. You can't flatten it perfectly unless you cut or tear the peel. But why should converting a flat surface into a round one, or a round surface into a flat one, be so difficult? What is it about round and flat surfaces that prevents them from transforming neatly back and forth into one another?

The answer lies within a slice of pizza and the writings of Carl Friedrich Gauss, a German mathematician who lived from 1777 to

1855. (Gauss has a special place in math history. Often regarded as one of the greatest mathematicians since the time of the ancient Greeks, Gauss is often referred to as the Prince of Mathematics. Remember that he also taught August Ferdinand Mobiüs—see entry #7) Gauss is responsible for a theorem about curved surfaces that's known as the *theorema egregium* (Latin for remarkable theorem).

To get an idea of what Gauss's theorem means, think of a human being who's been shrunk down to one inch tall and placed on the surface of a cylinder. If the person started walking around, he would find many kinds of paths he could take. For instance, he could walk across the top of the cylinder, in a straight path. Or he could walk around the curved side of the cylinder, going in a circle until he returned to his starting point. (We'd have to envision this tiny person wearing very sticky shoes.) He could also walk in a corkscrew fashion, circling the cylinder while simultaneously advancing along its length. Gauss's theorem maintains that you can measure the curve of that cylinder by taking into account all of these paths, multiplying them together, and getting a single value. The flat path has a curvature of zero—it's flat, after all—and the curved path has a positive curvature. (A concave curve—one that curves inward—would have a negative curvature.) When you multiply the curvatures together, you end up multiplying a positive by a zero, leaving you with zero (since anything multiplied by zero equals zero). The cylinder would then be said to have a Gaussian curvature of zero.

Gauss's theorem also has consequences for how a surface can be manipulated. It maintains that as long as you don't rip it, you can bend and stretch a surface and it will keep the same Gaussian curvature throughout. So no matter how you crumple or distort the cylinder, its Gaussian curvature will never change.

That brings us to pizza. If you've tried to hold a large slice flat, especially if it's oozing with cheese and pepperoni, you know that the tip flops down, making eating the slice difficult. On the other hand, if you fold the slice lengthwise, the tip now protrudes straight out in a line and your toppings stay where they belong. What is going on? Well, if you calculate the Gaussian curvature of an unbent slice of pizza, you get zero. (All the possible paths that a 1-inch person could take on the surface of the slice are flat.) That means that no matter how you move or bend that slice, it will retain its zero value.

Now examine the flopped-over pizza slice again. Notice that the path from crust to tip is curved, while the path from side to side is flat. But if we fold the slice, we are making the side-to-side path curved, meaning that the crust-to-tip path must now be straight.

What does all of this mean? No matter how the slice is bent, one possible path on its surface must be flat (because flat curves have a Gaussian curvature of zero, and we need a zero in our calculations to make sure that the result is also zero). If the side-to-side direction is flat, then the front-to-back direction must be curved. If the side-to-side direction is curved, then the front-to-back direction must be flat.

Back to our original problem: The reason why you can't neatly wrap a watermelon with a sheet of wrapping paper, or completely flatten a grapefruit peel, is that flat and rounded objects have different Gaussian curvatures. So the next time you order a pizza, remember Gauss and his theorem and fold your slices with confidence.

CARL GAUSS

Carl Gauss was a child prodigy. Once, while in school, he was asked to add all the numbers from 1 to 100. He reportedly came up with the solution in seconds, noticing that the sum could be broken down into 50 pairs of numbers—1 and 100, 2 and 99, 3 and 98, etc.—each adding up to 101. The result therefore is 101 × 50, or 5,050.

23

Geodesic Domes

MATHEMATICAL CONCEPT: GEODESIC DOMES

HAVE YOU EVER been to Epcot and stood beneath the gigantic sphere known as Spaceship Earth? If so, you've been in the presence of a structure called a geodesic dome. Geodesic domes are composed of triangular pieces placed next to each other in such a way that the top of one triangle is next to the base of its neighbor. So instead of a smooth sphere or dome, a geodesic dome is rough and slightly angular (kind of like a disco ball).

Popularized in the mid-1900s by Buckminster Fuller, a pioneering thinker interested in using engineering to solve humanity's problems, geodesic spheres and domes (which are just spheres cut in half) are extraordinarily light and strong. Building a structure of triangular pieces results in greater stability than building it of squares. Fuller envisioned the geodesic dome as efficient, affordable housing, an inexpensive dwelling for Earth's multitudes. The savings stem from the shape: spheres enclose a given amount of space with a minimum amount of surface area, in theory lowering the costs of building materials. The open interior space also allows for the easy movement of air, potentially reducing heating and cooling costs. In fact, geodesic domes are the embodiment of Fuller's maxim of doing more with less.

You can also think of geodesic domes as varieties of Platonic solids (see entry #20). Like those venerable and beautiful shapes, geodesic domes are composed of one kind of polygon—triangles—except that the triangles in geodesic domes are pushed out, in a way, so they are closer to the skin of an imaginary sphere encapsulating the dome. And as with Platonic solids, geodesic domes showcase the power and majesty of geometry.

If you want to see a geodesic dome in person, you don't have to go to Disney World. You can go to the Missouri Botanical Garden in St. Louis, and visit the Climatron, a dome that's seventy feet high and 175 feet in diameter, and is built with aluminum poles and Plexiglas panels.

Why aren't there more dome-shaped homes? It may have to do with the fact that domes have less usable space than typical homes do.

BUCKMINSTER FULLER

Buckminster Fuller was ahead of his time. Known for innovative inventions like the Dymaxion car, which had three wheels, Fuller was devoted to helping humanity and finding ways to do more with less. He also made contributions to language, coining terms like "Spaceship Earth" and "synergetic."

24

A Fictional Math Book?
Yes.

MATHEMATICAL CONCEPTS: GEOMETRY, DIMENSIONS

IMAGINE A WORLD of two dimensions filled with sentient shapes, squares and hexagons, lines and circles that think and interact much as we do in our three-dimensional world. That is the premise of *Flatland: A Romance of Many Dimensions*, a story published in 1884 by Edwin A. Abbott, an English schoolteacher and clergyman. The protagonist is a square, who tells the reader of Flatland's laws and customs, including the shapes of houses—pentagons, to make sure that the angles are not sharp enough to harm Flatland inhabitants who might unwittingly run into them—and the hierarchy of its inhabitants. Women in Flatland are straight lines, while soldiers and lower-class workmen are isosceles triangles that have one extremely acute angle (all the better for warfare). Middle-class males of this world are squares and pentagons (five-sided shapes), while nobles are hexagons (six sides). As the number of sides increases, so does the rank of that shape; the highest rank is occupied by circles, who enjoy a kind of priestly caste.

The remarkable aspect of the story is how well it explains the concept of dimension. The square travels to Lineland and Pointland, and interacts with inhabitants of Spaceland, a place of three dimensions that the square tries to understand. And you can imagine how difficult it would be to try to explain the third dimension to a two-dimensional creature. Indeed, how would you do so? You could try to tell the creature that the third dimension is "up," or perpendicular to the flat world that it inhabits, but what would that mean to the creature? How could that creature envision a direction that does not lie on a flat plane, that instead rises away from it in some unknown fashion? *Flatland* helps the reader understand the nature of dimension in a way that has not been surpassed since the book's publication.

FLATLAND: THE MOVIE

If you don't feel like reading the book, you can always watch *Flatland: The Movie*. Released in 2007, it features the voices of Martin Sheen, Kristen Bell, and Michael York, and focuses on the story of Arthur Square's journey.

25

A Soccer Ball Is More Than Just a Ball

MATHEMATICAL CONCEPTS: SHAPES, GEOMETRY

THAT SOCCER BALL you kick around on the weekends has some mathematical secrets. If you look closely, you'll see that the ball is covered with pentagons (five-sided shapes) and hexagons (six-sided shapes) in a repeating pattern. In fact, this pattern of shapes means that soccer balls are truncated icosahedrons: they have 12 pentagonal and 20 hexagonal faces, for a total of 32. Moreover, each side of each pentagon touches a hexagon, while the sides of each hexagon alternate between touching a pentagon and another hexagon. In a pure truncated icosahedron, however, the pentagons and hexagons are completely flat. The pentagons and hexagons on the soccer ball are puffed out to eliminate the edges and make the ball round.

The truncated icosahedron stands out among other shapes because it's one of the Archimedean solids. Named for Archimedes, one of the greatest mathematicians of ancient Greece (and all of human history), these three-dimensional forms have faces consisting of two or more kinds of regular polygons (shapes whose sides all have the same length, such as hexagons). A related form, the Platonic solid (see entry #20) has faces that are all the same kind of regular polygon. (Think of a cube, whose faces are all squares.)

Truncated icosahedra are not just found in the world of sports. They also appear in nature, at the microscopic level, in the form of a buckminsterfullerene, a molecule made up of sixty carbon atoms. Named after Buckminster Fuller, one of the twentieth century's most iconoclastic thinkers and engineers, and only discovered in the 1980s, the molecule occurs in the shape of a ball. Some viruses, like the cowpea chlorotic mottle virus, have the shape of a truncated icosahedron. This special form seems to occur throughout both the natural and manmade worlds.

OTHER ARCHIMEDEAN SOLIDS

There are thirteen Archimedean solids. Some of the others include the cuboctahedron, which has faces that are squares and triangles, and the truncated octahedron, whose faces are squares and hexagons.

26

Rubik's Cube: A Toy, or a Mathematical Wonder?

MATHEMATICAL CONCEPTS: SHAPES, COMBINATORICS, ALGORITHMS

YOU MAY NOT have played with one lately, but the Rubik's Cube, that multicolored puzzle introduced to the world in the early 1980s, is considered the bestselling toy in history. With six faces, each composed of three movable layers and nine minicubes, with each minicube having one of six colors, the Cube offers a harrowing and addicting challenge: once the Cube has been scrambled, puzzle-solvers must rotate the layers until each side of it is all one color.

Though both maddening and fun, the Rubik's Cube also taps into several veins of mathematical thought. One is combinatorics, which deals with the various ways in which a set of entities can be ordered. There are an astounding number of ways in which the Rubik's Cube's minicubes can be arranged. In fact, the total number of permutations (another word for arrangements) is 43,252,003,274,489,856,000, or around 43 quintillion. It's profoundly difficult to grasp how large this number is. To start, consider that if you had 43 quintillion Rubik's Cubes and stacked them one on top of another, the tower of Cubes would reach into outer space. How far exactly would they extend? To the International Space Station? To the moon? The answer might surprise you: once the tower had been built, it would extend for *261 light years*.

You can try to grasp this mind-boggling number in other ways. For instance, 43 quintillion Rubik's Cubes would be enough to cover the entire surface of Earth, not once, not twice, but *273 times*. Or, think about how long it would take you to click through all of those permutations. If you could turn a layer of a Cube once a second, going through every possible arrangement would take you almost 1.5 quadrillion years, far longer than the age of the universe.

Pondering the nature of the Rubik's Cube is also a good way to learn about algorithms. In math, an algorithm is a set of instructions that take you from one state of affairs to another, through a series of specific steps. (You can think of the instructions to put together your new IKEA bookshelf as an algorithm.) People who compete to solve the Rubik's Cube in the shortest amount of time—a kind of puzzle sport known as speedcubing—memorize algorithms that instruct them how to twist the Cube's layers to move certain mini-cubes into position. The algorithms use a notation in which a letter or symbol stands for one of the Cube's faces. For instance:

F = front face
B = back face
R = right face
L = left face
U = top face
D = down face

There are also symbols that tell the puzzle-solver whether to twist the face clockwise or counterclockwise. For instance, F would tell someone to twist the front face clockwise; F' would stand for a counterclockwise turn. So, an algorithm used to solve the middle layer of a Rubik's Cube might look like this:

U R U' R' U' F' U F

Speedcubers regularly memorize up to forty of these kinds of algorithms.

THE WORLD'S BESTSELLING TOY

An estimated 350 million Rubik's Cubes have been sold throughout the world since 1980, making it the world's bestselling toy. That means that approximately one in seven people have played with a Rubik's Cube.

27

Paper Sizes

MATHEMATICAL CONCEPTS: GEOMETRY, RATIOS

THE NEXT TIME you use a photocopier, give the sheets of paper a second look: a great deal of mathematical planning has gone into designing them. The dimensions specified by the International Organization for Standardization (ISO), especially for its A and B series of paper, were formulated with a particular geometry in mind, one that has advantages when you're making copies.

A special property of A series and B series paper is the ratio of a given sheet's two edges. For both categories, the ratio of the width to the height is 1:$\sqrt{2}$. This property means that each sheet of A4 has one-half the area of an A3 sheet, and each sheet of A3 paper has one-half the area of an A2 sheet. Using $\sqrt{2}$ means that each size of paper has the same width-to-height ratio; each one is a perfectly scaled version of the next-highest or next-lowest size. For example, the A4 paper size—which roughly corresponds to America's letter paper size—has a width of 210 millimeters and a height of 297 millimeters. The corresponding dimensions of an A3 sheet of paper are 297 and 420 millimeters.

As a result, were you using a photocopier and wanting to reduce the size of an A4 sheet, you could convert it to the A5 size, which when rotated would give you two copies that fit *exactly* on one A4 sheet, with no wasted space. And because each size has the same ratio, no matter how much you scale up or down, the information on the paper will appear in the same aspect ratio. Even in a task as mundane as photocopying, geometry can make our lives easier.

WHAT IS A QUIRE?

The terms found in the world of paper are often unfamiliar to the average person. For example, a quire is a set of 24 or 25 sheets of paper that are all the same size, equaling $\frac{1}{20}$ of a ream (which is 400 or 500 sheets of paper).

28

Different Ways to Depict Earth on a Map

MATHEMATICAL CONCEPTS: STEREOGRAPHIC PROJECTION, MERCATOR PROJECTION, ROBINSON PROJECTION

IF YOU'VE EVER looked at a map on someone's wall or in a road atlas, you were looking at math at work. As you learned in entry #22, when you read about Gauss and pizza, it's impossible to perfectly convert a spherical shape into a two-dimensional shape. As a result, any map of Earth—or any other planet or spherical body—will include some distortion. But how exactly do you translate the information on a sphere into information on a sheet of paper? In other words, how do you convert a globe into a map?

That's where math comes in. There are different kinds of maps, and each depicts Earth in a slightly different way. Each variety is called a projection. You might have already heard of the Mercator projection (presented by Gerardus Mercator, a Flemish cartographer, in 1569), which was of great use to sailors, since to get from point A to point B, all the navigator had to do was draw a line from one to the other and he would know the exact compass bearing that would get him to his destination. And if you've ever seen a wall map of the world published by *National Geographic*, then you've seen a Robinson projection. (Robinson maps were developed to show less distortion in the polar regions, which in Mercator maps look vastly larger than they are in life.)

Some maps are shaped like circles and are often centered on Earth's North or South Pole. You might have seen antique maps in this configuration, with two circular maps set together in a frame. This kind of map is a stereographic projection. Unlike some maps, stereographic projections are conformal; that is, all the angles in them are true to life. (The distances and areas, however, are not.) In addition, circles that go around Earth's globe are depicted as circles in a stereographic map. If the circles happen to pass through the point of projection—the center of the map—then they will appear in the map as straight lines.

THE GALL-PETERS PROJECTION

In some projections, including the Mercator, the relative sizes of the continents are distorted. The Gall–Peters projection, named for James Gall and Arno Peters, attempts to correct some of the distortion, making the relative sizes of the continents more accurate. You may remember that it was championed in a famous episode of *The West Wing* by the fictional Organization of Cartographers for Social Equality.

29

Packing M&M's

MATHEMATICAL CONCEPT: COMBINATORICS

MATH MAY SEEM esoteric and disconnected from everyday life, but you can find it in the most humdrum of places. In fact, a link to seventeenth-century mathematics might be no farther than the candy aisle at the nearest grocery store.

In 1611, Johannes Kepler—famous for devising laws of planetary motion (see entry #20)—hypothesized that if you are using particles shaped like spheres, there is no better way to pack a space than by organizing the spheres like a stack of oranges at a market. This is known as the Kepler conjecture.) Using this technique, called face-centered cubic packing, a person can fill approximately 74% of a given volume. If the spheres are dumped into a jar haphazardly, though, they will fill only approximately 64% of the volume.

Enter the candy. Researchers have found that particles that look like M&M's—squashed spheres, or spheroids—pack into a jar just as well as spheres do. If they are stacked like oranges, they, too, fill approximately 74% of the volume. But if they are randomly poured into the jar, they beat the spheres, filling approximately 71% of the space, much *more* than spheres do. Some people think that spheroids can fill a volume more efficiently than spheres because they can turn and tumble until they pop into a configuration that uses up more space. Other shapes can do even better. Randomly packed ellipsoids—which are similar to the shape of footballs, or, if you like, candy-coated almonds—can fill up to 74% of a given volume.

Kepler was never able to prove his conjecture, though Gauss was able to construct a limited proof in the 1800s. The last step in establishing that Kepler was correct occurred in the 1990s, when mathematician Thomas Hales used a computer program to help construct a proof. The proof was so long, however—several hundred pages, in fact—that he used a computer algorithm to check it!

BLUE M&M'S

In 1995, candy aficionados voted on a new color to be added to the M&M's bag. The winning color was blue, which received 54% of the vote. (More than 10 million votes were cast.) Runners-up included pink and purple. The new blue candies replaced the tan plain M&M's.

30

Tangrams

MATHEMATICAL CONCEPTS: SHAPES, GEOMETRY

IF YOU LIKE GAMES, you may be a fan of tangrams, a puzzle from China that some people believe to have originated thousands of years ago, though the first published documentation was in 1813. The classic tangram set consists of seven shapes: two large triangles, one medium triangle, two small triangles, one square, and one parallelogram (a rectangle in which the two short sides are slanting in the same direction). All the triangles are right triangles, meaning that of the three angles in each triangle, one measures 90 degrees. The shapes can be made of almost any material, including wood, plastic, glass, or tortoiseshell. In fact, you can make your own tangram set using nothing but paper, a pencil, a ruler, and a pair of scissors.

The goal of the game is to arrange the seven pieces to form complex shapes, like animals and people. (Tangram sets often come with a book of suggested shapes.) The pieces cannot overlap, and an edge of each piece must touch the edge of at least one other piece.

The connection between tangrams and mathematics is clear: the shapes are found in geometry, the branch of mathematics that deals with lines, points, and angles. But tangrams have also inspired deeper mathematical reflection. Some mathematicians have wondered how many shapes one can create using the seven pieces in a tangram set. But the shapes they had in mind weren't those of sheep or sailors. They were instead convex polygons, shapes like pentagons and squares that have three or more sides and no places where the sides dip in toward the center. Mathematicians have discovered that a player can use all seven tangram pieces to form thirteen convex polygons: two pentagons (five sides), six quadrangles (four sides), one triangle, and four hexagons (six sides). The puzzle is simple but, as with much of mathematics, has a profound aspect that's not immediately obvious.

EGG OF COLUMBUS

Not all tangram sets are made up of triangles and rectangles. One variety, the Egg of Columbus, begins as a two-dimensional egg shape. It's then divided into a series of pieces, some of which have curved edges.

31

The Velvet Rope as a Mathematical Entity

MATHEMATICAL CONCEPT: CATENARY CURVES

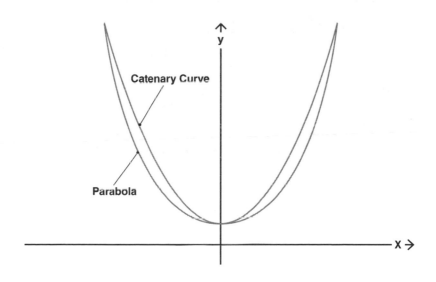

IF YOU VISIT ST. LOUIS, MISSOURI, you won't be able to miss the Gateway Arch, an enormous steel-and-concrete structure towering 630 feet above the ground. Completed in 1965, the arch was meant to symbolize St. Louis's historic role as a gateway to the West for the pioneers who colonized North America. The arch can also be seen as an homage to mathematics, because its shape closely matches that of a catenary, a kind of arch that forms when a chain is fastened to posts at both ends and allowed to drape toward the ground. (More accurately, the Gateway Arch is an upside-down version of an almost-catenary.) You can see catenaries in the shape of a power line hanging between transmission towers, and the shape of a heavy cable securing a ship to its port. You'll also find them in the velvet ropes that cordon off lines of people waiting to get into a movie or concert.

Catenaries look similar to parabolas, another kind of curve, but the equation that describes them was found only in 1691—a recent date in the history of mathematics—by a trio of science titans: Christiaan Huygens, Jakob Bernoulli, and Gottfried Leibniz.

CATENARY CURVES IN ARCHITECTURE

Upside-down catenary curves show up frequently in architecture, providing beauty and grace to all kinds of spaces. They appear under the terrace in Antoni Gaudí's Casa Milà, for instance, and support the roof of the Sheffield Winter Garden in South Yorkshire, UK.

32

How Do Suspension Bridges Hold Cars?

MATHEMATICAL CONCEPTS: SHAPES, PHYSICS

IMAGINE A BEAUTIFUL piece of architecture and you're likely to think of a suspension bridge. These soaring, swooping structures can be recognized by their cables, which are not only pretty but also serve a purpose: they support the roadway that passes below them. These undulating curves also happen to be examples of parabolas, a shape familiar to mathematicians and found in many places in the physical world.

If you remember the Cartesian coordinate system from geometry class, you also might remember that you can graph a parabola using the function $y = x^2$. Or, you might recall that the parabola is one of the class of shapes known as conic sections, which are formed by passing a plane, like a sheet of paper, through a cone in various ways. (You might have also learned in physics class that the cables transfer the downward force exerted by the heavy road and the cars on it to the bridge's towers, which then direct that force down into the earth.) Also, the cables are in the shape of a parabola partly *because* of the road and traffic going across it. Without that weight, the cables might take a shape closer to a catenary, which is the shape a hanging object like a rope takes when the only force acting on it is gravity (see entry #31).

TRUSS BRIDGES AND TRIANGLES

Not all bridges have cables, of course. Others are built using structural components known as trusses. Trusses are often built of many components, usually triangles, which connect so that the entire structure behaves as a unit.

PART 2
BEHAVIOR

33

Why Do Buses Arrive in Bunches?

MATHEMATICAL CONCEPT: CHAOS THEORY

IF YOU LIVE SOMEWHERE with a bus system, you've probably found yourself waiting at a bus stop for much longer than you had hoped, with no buses in sight. As you scan the horizon, your toe tapping with impatience, you suddenly see two buses coming, bunched together. "Arrrgh," you yell. "Why can't the buses stay spread out? What's wrong with the transportation system in this stupid city?"

It turns out that buses don't bunch up because there's something wrong with the transit system. In fact, bunching is almost inevitable. Consider two buses that leave the bus depot early in the morning, at the very beginning of their runs. And let's say that the buses leave the depot at ten-minute intervals, so they begin the day spaced ten minutes apart. If all a bus had to do was arrive at a bus stop, wait an allotted amount of time, and then drive off, bunching might never occur. But, of course, people have to board the bus at each stop, and the amount of time it takes each person to board a bus varies. (Compare an elderly man using a walker to a ten-year-old boy.) Moreover, some bus stops might get flooded with waiting passengers. (Maybe one stop is near a school, and every day at 3:00 P.M. crowds of students pour out and wait at nearby stops to go home.) In either case, a bus might get bogged down at any point along its route.

If you think about this bogging down, the way that buses bunch up becomes a little clearer. When one bus has to wait a long time at a bus stop to let passengers on, whether because some individuals board more slowly or because the overall group is large, there is more time for more people to accumulate at the next stop on the route. Then, when the bus finally moves on and arrives at that next stop, those accumulated people take even *longer* to board the bus, meaning that an even larger crowd will be gathering at the *next* stop. This slowing-down process, in effect, builds on itself.

Meanwhile, the bus behind the slow bus is finding that when it arrives at a bus stop, there aren't many people waiting for it. That's because most of them have gotten on the previous bus (the slow one). Since there are fewer waiting passengers, the bus doesn't have to wait very long for people to board. So it can leave the bus stop relatively quickly, shortening the time it takes to get to the next stop. And just as the slow bus gets slower and slower, the fast bus gets faster and faster. Eventually, the fast bus catches up to the slow bus and the two continue along the route together (unless the slow bus starts to skip stops to increase the spacing).

Bus clumping is an example of chaos theory, a branch of mathematics that explores how small differences in initial conditions can lead to dramatic differences in end conditions. In this case, tiny variations in how long people take to board a bus dramatically influence a bus's position relative to all the other buses on its route.

BUTTERFLY EFFECT

If you've heard anything about chaos theory, then you've probably also heard about the butterfly effect. It was first described in the 1960s by mathematician Edward Lorenz, who was studying weather patterns when he noticed that slight variations in data plugged into a weather prediction model resulted in vastly different outcomes. According to the butterfly effect, a very small weather input, like the flapping of a butterfly's wings, can lead to a massive disturbance, like a hurricane.

34

Stop Losing Money at the Casino

MATHEMATICAL CONCEPT: GAMBLER'S FALLACY

DO YOU LIKE TO GAMBLE? If so, you may have been inter-acting with math without knowing it, in the form of the gambler's fallacy.

Let's imagine you are throwing dice. Each die has six numbers—1, 2, 3, 4, 5, 6—so the probability that any of those numbers will come up after a throw is one in six. Let's say that you roll the die ten times. Let's also say that these are the results you get:

Roll 1: 5	Roll 6: 3
Roll 2: 2	Roll 7: 4
Roll 3: 1	Roll 8: 5
Roll 4: 2	Roll 9: 3
Roll 5: 4	Roll 10: 3

People often believe that because a 6 hasn't come up yet, the odds of its coming up next are higher than they would have been if 6 had come up, say, three times during the previous ten rolls. "I haven't rolled a 6 yet," someone might think to himself, "so it's due to come up any roll now." But in fact, the odds of the 6 coming up are no different than they were at the first roll. The fallacy, or error in reasoning, lies in thinking that previous rolls influence future rolls. Even if you rolled the die 100 times and a 6 never came up, the odds of a 6 coming up on roll 101 would still be one in six.

The gambler's fallacy goes by another name, the Monte Carlo fallacy, which probably stems from one evening in a Monte Carlo casino in August of 1913. On that night, black came up on a roulette wheel twenty-six times in a row. In the spins leading up to the fateful twenty-seventh, more and more gamblers bet on red, sure that red was more and more likely to come up the longer the streak continued. Of course, the odds of either black or red coming up don't change from one spin to another. As far as the odds are concerned, each spin is the first.

Mathematicians and statisticians sometimes describe the gambler's fallacy as a belief that inanimate objects—like roulette wheels or dice—have memories. This belief extends to the notion that "knowing" how they have behaved in the past, these inanimate objects adjust their future performances accordingly. But of course, those objects do not have memories; each successive spin of the wheel or roll of the die is completely independent of any other (assuming that the equipment hasn't been tampered with). So keep that in mind when you're deciding what to bet on!

FALLACIES

Another example of a fallacy might be the ad hominem argument, in which a person's character is offered as evidence for or against a proposition, when in reality a person's character has nothing to do with the proposition's truth value. For example, let's say that you and another person were in a debate about the death penalty, and you were each trying to persuade a mutual friend that your respective viewpoints were correct. Let's also say that your opponent in this debate had recently cheated on his wife. If you then said to the third party, "You shouldn't take my opponent seriously; after all, he committed adultery," you would be committing an ad hominem fallacy. Your opponent's personal life has nothing to do with the validity of his arguments.

35

How Does a Movie Win an Oscar?

MATHEMATICAL CONCEPT: COMBINATORICS

DURING THE AIRING of the Academy Awards, friends and family gather around the TV and wait until the end of the telecast to find out which movie won the Best Picture award. But how exactly does the Academy determine which movie wins? It turns out that mathematics has a say in this process, in the form of a procedure known as preferential voting.

In the kind of voting most people are familiar with, you get a ballot with a list of candidates and you place a mark by the name of the person you would like to win. (Let's pretend this is an election for your state's governor.) An official then reads over the ballots, counts the number of marks for each name, and determines who got the most. That person then becomes governor.

In preferential voting, voters put a mark beside every name on the ballot, not just the one they want to win. If the ballot has five names on it, the voter would put a 1 by the person he most wants to win, then a 2 by the person he *next* most wants to win, and so on, down to 5. When the election official sits down to examine the ballots, he makes piles for each of the five candidates. All the ballots with a 1 by candidate A's name get placed in candidate A's pile, all the ballots with a 1 by candidate B's name get placed in candidate B's pile, and so on. At the end, if one of the piles contains more than 50% of the 1s, then that person wins. If not, the candidate with the least number of 1s is removed from consideration. But the ballots live on! The election official goes through all the ballots again, looking at which names have a 2 by them, and the process repeats itself. All the ballots with a 2 by candidate A's name are placed in the candidate A pile, and so on. Like before, with the 1s, the candidate who had the least number of 2s placed by her name is removed from consideration. This process continues with 3s, 4s, and higher numbers until some candidate has more than 50% of the ballots in her pile. Voting is not always as simple as it seems.

OSCARS BY NUMBERS

Numbers rule the Oscars. Since the first Academy Awards ceremony in 1929, more than 3,000 Oscar statuettes have been handed out. The event's sponsor, the Academy of Motion Picture Arts and Sciences, has roughly 6,000 members. And the movies with the most nominations are *All about Eve* and *Titanic*, each with fourteen.

36

Staying Dry in the Rain

MATHEMATICAL CONCEPTS: SHAPES, ARITHMETIC

ONE CLASSIC PLACE where mathematics can appear is in the middle of a rainstorm. Let's say you're caught in a downpour without an umbrella: What should you do to minimize how wet you get? Standing still, of course, is not an option. You'd just get soaked. Your two realistic options, it seems, are to walk or run to the nearest shelter. If you walk, it seems like you'd get wetter because you'd be out in the rain longer. If you run, it seems like you'd get wetter because you'd collide with more raindrops as you ran. What's the answer?

Math can help us figure that out. To begin, let's reformulate the problem so it's a little easier to handle. First, instead of a real person, imagine an idealized person made up of a three-dimensional rectangle (like a giant brick). Second, imagine that the rain is falling at a constant rate: there aren't any sudden, intense downpours; neither are there any breaks in the rain. Third, let's stipulate that the rain is falling straight down, at a 90-degree angle to the earth. Now we have a nice, simple scenario that we can wrap our brains around.

Let's determine how much rain—the volume of rain, to be specific—will fall on the brick person's head (which is a nice flat surface). We know that since the rain is falling straight down at a constant rate, as the brick person moves forward, either walking or running, the rain will fall on the top surface at a constant rate. This constant rate has a surprising consequence: no matter whether the person walks or runs, he will move beneath an identical column of rain. You can imagine these rain volumes as three-dimensional rectangles, too: the rectangle of rain that hits a person standing still would look like a standard, straight-up-and-down rectangle; the rectangle of rain that hits a person walking or running forward would look slanted. But—and here's the crucial part—the volumes of the straight rectangle and slanted rectangle are the same. (You calculate the volume of a 3D parallelogram, more properly called a parallelepiped, by multiplying length times height times depth.) Similarly, because the surface area of a person's front is always the same, as is the rate of the falling rain, the volume of water that a person hits while either running or walking is the same.

If we wanted to represent the total volume of water that falls on a person caught in a rainstorm, we could write the following equation:

Total Volume − [*time spent in rain* × *rate of rainfall*] + [*distance to shelter* × *rate of rainfall*]

Because the distance between you and the shelter doesn't change, the only way to minimize how wet you get is by spending as little time in the rain as possible. The only way to do *that* is to run as fast as you can.

COUNTERPOINT BY ALESSANDRO DE ANGELIS

Alessandro De Angelis, a physicist at the University of Udine in Italy, determined that if you ran, rather than walked, through a rainstorm, with all variables remaining constant, you would stay only 10% drier. So in a 1987 paper published in the *European Journal of Physics*, he concluded that it was better to walk, since the difference wasn't significant enough to exert the extra energy.

37

The Most Efficient Checkout Line

MATHEMATICAL CONCEPT: QUEUING THEORY

SHOPPING FOR GROCERIES can be fraught with all kinds of frustrations. Someone's cart might be blocking an aisle. Your favorite kind of cereal could be sold out. And where is the hummus?

But perhaps the worst frustration, the one that can sink into your soul and fester, is the act of waiting in the checkout line. There you are, cart full of Rice Chex and spaghetti and apples, faced with the task of choosing which checkout line is moving the quickest. But just as you make your choice, your line seems to slow down, thanks to a shopper who is fumbling with coupons or loose change. Now every other line seems to be moving faster than yours. Why does it never seem like the line you choose is the fastest?

There happens to be a branch of mathematics that deals with just this issue. It's called queuing theory, and it deals with the behavior of waiting lines. (Mathematicians who specialize in queuing theory are called queuing theorists.) The field originated in Copenhagen in the first decade of the twentieth century. An engineer and mathematician, Agner Krarup Erlang, was trying to figure out the minimum number of phone lines the city would need to make sure that most phone calls would go through. (At this time in history, calls were connected by people, who had to plug a jack into a hole for each call.) The phone company wanted to avoid having either too few lines, which could cause a backup if lots of people wanted to make calls at the same time, or too many lines, which would mean that the company had paid for equipment it didn't need.

Erlang's name is forever tied to telephony: an erlang is a unit of telephone load, or telecommunications traffic, and is used to measure traffic volume. And his findings have applications outside of telephone networks, including traffic engineering, the Internet, and how factories are designed.

But you probably have encountered queuing theory yourself while doing your errands. Queuing theorists have found that if customers form one long winding line, called a serpentine line, and then are sent to the next available register, wait times can be drastically reduced. (Serpentine lines can be found at banks, where people wait for the next available teller, or at some grocery stores.) Serpentine lines ensure that wait times are minimized because, instead of the traditional-line method, in which one slow person or teller can delay an entire line, a slow person can tie up a register but meanwhile the other customers can be shunted to other open registers. The delay remains, but its impact is much lower that it would be otherwise.

LEFT OR RIGHT?

When faced with two choices—a line to the left or a line to the right, some people believe that the left-hand route will be faster. That's because approximately 90% of the population is right-handed, and so they tend to naturally head to the right. This may be an old wives' tale, but if you're at a theme park with long lines, heading left is worth a try.

38

How to Study for the Turing Test

MATHEMATICAL CONCEPT: TURING TEST

IF YOU'VE SEEN the 1982 movie *Blade Runner*, you know the scene: a man sits at a desk and through a haze of cigarette smoke tries to determine whether another man, seated on the other side of the desk, is a robot. This idea—a test for consciousness—seems like something only found in twentieth-century science fiction, but in truth it has been around for centuries. René Descartes mentions it in his 1637 book *Discourse on Method*, where he argues that if there were a machine that looked and acted like a human being, one would nevertheless be able to tell that it was artificial.

1. One, the machine would not be able to speak convincingly in many kinds of situations—it would, in other words, never be able to progress beyond preprogrammed speech.

2. Two, it would never be able to act in a universal way. (Descartes means that machines tend to be specialized and good only at particular tasks, like welding or printing.

Because their parts are designed for a limited number of purposes, machines don't have the ability to interact with the world creatively and spontaneously.)

The most explicit example, though, of a procedure that would distinguish a machine from a thinking creature was presented in a 1950 paper by Alan Turing, a British mathematician and cryptologist who during World War II helped the Allies crack the Nazi's Enigma code. His paper, titled "Computing Machinery and Intelligence," posited a test that might help settle the question of whether a machine could ever be said to think.

Because it's difficult to define exactly what thought is, or what thinking involves, Turing proposed a different way to attack the problem. His test, originally known as the Imitation Game, asks instead whether a machine could ever fool a person into believing that it was a human being. In the game, a human interrogator sits in one room, while in two other rooms are a machine (let's say it's a computer) and another human. The interrogator can send messages in text form to both the machine and the human, and both machine and human can send responses in return. The interrogator's task is to figure out which is the machine and which is the person. If he cannot do so, or if he is fooled one-third of the time, the machine passes the test. According to Turing, if the machine passes the test, it's reasonable to claim that the machine is intelligent. (After all, isn't the ability to converse an important way we judge that other human beings are intelligent?)

The Turing Test is fascinating because it challenges what most people believe the process of thinking entails. Most people would say that thinking is something that happens *inside* someone's brain, a hidden drama forever invisible to other people. But the Turing Test suggests that we don't have to have access to someone's interior world to find out whether a mind lives there.

The questions asked during a Turing Test don't have to be especially probing or complicated. Mundane, or even boring, questions will do just fine. For instance, in a Turing Test recently conducted in the UK's Reading University, the machine was asked what the weather was that day, what its favorite subject in school was, and whether it liked soccer.

Still, the test raises almost more questions than it answers. Does passing the test really indicate intelligence, or does it merely show that a computer program has succeeded in imitating a human? Does it prove there's more going on in the program than a purely mechanical shuffling of symbols? And if we assert that all that's happening is the passing to and fro of electrons, without anything mind-like behind all of that commotion, how do we know that something similar isn't happening inside a person's brain?

Even today, the Turing Test has not lost its appeal. Turing Tests are administered every year; in 2014, a Russian chatbot named Eugene Goostman "passed," having convinced 33% of the judges that it was human. Some people dispute the victory, though: Eugene was designed to mimic a thirteen-year-old Ukrainian who had learned English as a second language.

Despite Turing's achievements, both in wartime and peacetime, Britain treated him cruelly. Having learned that Turing was homosexual, and since homosexuality was at that time a criminal offense, the British government arrested him in 1952. Because it was thought that homosexuals could be blackmailed, Turing's security clearance was revoked, and he was offered the choice of imprisonment or forced injections of estrogen, to extinguish his sexual drive. He chose the injections. Perhaps as a result of the way the government treated him, in 1954 Turing committed suicide.

The Turing Test showed the possibilities of a digital computer, one that can perform mathematical operations using purely mechanical means (see entry #70). Digital computers are of course the ancestors of today's laptops and smartphones, which can not only add, subtract, and multiply, but also run complex programs like Facebook and all kinds of web browsers. Turing and his ideas, therefore, helped create the field of artificial intelligence, which resonates in science fiction and in engineering departments across the nation.

THE IMITATION GAME

Alan Turing has recently entered the public consciousness in a new way. A movie about him was released in 2014, called *The Imitation Game*. It stars Benedict Cumberbatch as Turing and chronicles the race to crack Germany's Enigma code during World War II. The movie chronicles how Turing helped crack the code by using a precursor to a modern computer. The specific techniques he used recently came to light due to the release of declassified documents.

39

What Is a Sextant?

MATHEMATICAL CONCEPT: GEOMETRY

IF YOU WERE SAILING a boat on the open ocean, how would you determine where you were? To make the task more difficult, let's rule out using GPS or anything that relies on electricity. (Google Maps is not allowed.) The problem sounds insurmountable, but sailors have been figuring out their locations for centuries, so we know it can be done. What's the secret?

The answer relies on angles and geometry. To begin, let's review what makes this kind of navigation possible. If you've seen a globe, then you have already seen the lines that crisscross it. Some are horizontal, extending above and below the equator (the line that encircles the globe in the very middle). These lines are known as lines of latitude. Other lines are vertical; they run north and south and intersect at the North and South Poles. These are known as lines of longitude. Knowing your longitude requires a clock, but what we're interested in is knowing how to find one's latitude. That is the navigational knot that is untied using mathematics.

A crucial key is realizing that the position of the sun at noon on any particular day of the year depends on your present latitude. The closer you are to the equator, the more the noonday sun

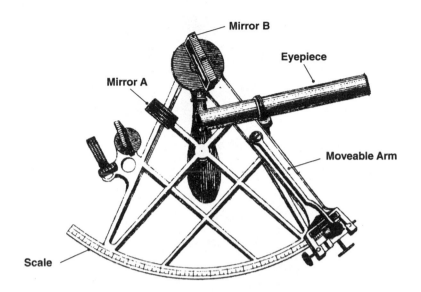

appears to be 90 degrees above you. The further you move toward the North or South Pole, the more the angle decreases; in other words, as you travel north or south, the sun at noon appears lower and lower in the sky. You can also apply those principles in reverse. If you can figure out the angle of the noonday sun with respect to you, you can deduce your latitude.

Figuring out these angles is the job of the sextant, a handheld measuring instrument that looks like a metal slice of pie with some doodads attached to it. One of those doodads is a sighting telescope. To use the sextant, you look through the telescope at a celestial object, like the moon, a star, or the sun (through a filter, of course). The image appears on two mirrors. You then move the index arm, a piece of metal that slides along the edge of the curved pie piece, until the image of the celestial object in one of the mirrors touches the horizon. At that point, you look at the pie piece, which has angle markings on it. The index arm will be pointing to an angle. That angle can be used to determine your latitude.

For example, let's say that on June 21, 2015, you're on a boat off the coast of Christmas Island, an Australian territory near Indonesia in the Indian Ocean. Using your sextant, you'd be able to determine that the sun was 66.6 degrees above the horizon. With that information, you could determine that your latitude was 10.48 degrees south.

In essence, you are using trigonometry—the mathematical study of the properties of triangles, including their angles—to fix your position.

If you thought that geometry was irrelevant to your everyday life, think again, especially if you find yourself lost on a boat without electronics!

JOHN CAMPBELL

The first true sextant was invented by John Campbell in 1757 and was first used to its full potential, including as a device to tell the time, by explorer Captain Cook in 1768 when he set off to chart New Zealand.

40

Splitting Up Rent

MATHEMATICAL CONCEPTS: FAIR DIVISION, COMBINATORICS

IF YOU'VE EVER had roommates, you'll be familiar with the challenge facing three or more people who split the rent on a house or apartment. Figuring out how to fairly determine who pays how much can be more complicated than it seems. The task is difficult because rooms are often very different—for instance, some have more light while others have more space—and each person might value each aspect differently. How can the rooms and rent be allocated so that each person is happy with his share, while not being envious of any other housemate?

These problems fall under the category of fair division. A kind of problem that belongs to many realms including mathematics, economics, law, and politics, fair division is concerned with dividing up goods so that each party receives a just amount. The division should also be made in such a way that each party would not want to exchange her share of the goods with any other share. Examples of fair division pop up during divorces, auctions, and even war.

In 1999, Francis Su, a mathematics professor at Harvey Mudd College, published a paper explaining how to solve fair division problems using Sperner's lemma, a theorem that addresses a branch of math known as combinatorics (see entry #26). Originally, the lemma made a claim about triangles. Take a triangle and divide the interior of it into smaller triangles. You can create as many as you like; just make sure they all fit snugly together with no empty space between them. Next, label each of the larger triangle's vertices—the corners—1, 2, and 3, so each corner gets a different number. At this point, notice that corners of some small triangles touch at least one of the legs of the big triangle. At each of these points, write down a number. On the leg between corners 1 and 2, label each point either 1 or 2. (What number you assign to which point is up to you.) On the leg between corners 2 and 3, label each point either 2 or 3, and on the leg between corners 3 and 1, label each point either 3 or 1. As for the corners inside the big triangle, you can label those either 1, 2, or 3, in whichever way you choose. Sperner's lemma states that there must be at least one small triangle with vertices labeled 1, 2, and 3. There could be more than one, but there can only be an odd number of them, never an even number.

When Sperner's lemma is applied to the rent problem, the numbers are replaced with letters representing the first names of each tenant, and each triangle, both large and small, represents a

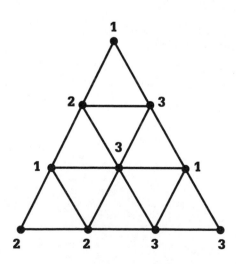

different allocation of rent. According to the lemma, there must exist an allocation of rent that satisfies each tenant such that no one is envious of anyone else's room and rent situation. In other words, because there is a small triangle within the large triangle that has vertices labeled 1, 2, and 3, there must also be a way to allocate rooms and rent so everyone is happy.

Sperner's lemma is an example of a mathematical finding that might seem abstract and irrelevant to everyday life, but in fact can help people solve problems effectively and efficiently.

FAIR DIVISION AFTER WORLD WAR II

A specific instance of fair division occurred after World War II, when the allies had to figure out what to do with Berlin. They ended up dividing the city into four sections. The Allies' three sections (ruled by the United States, the United Kingdom, and France) formed West Berlin, while the Soviet section formed East Berlin.

41

Cutting a Cake Fairly

MATHEMATICAL CONCEPT: FAIR DIVISION

THE NEXT TIME you're at someone's birthday party, consider that the simple act of cutting a cake has spawned a vast amount of mathematical thinking. How can one make sure that each person is satisfied with the slice she gets, and moreover does not want anyone else's slice more than her own? This goal becomes more complicated when one realizes that not everyone likes the same kind of cake slice: some like more frosting; some like less. Some want a flower on their piece; others want letters. The question that mathematicians have attempted to answer is whether there's a way to divide a cake so that each person who gets a slice is happy. In fact, an ideal method of cutting a cake between two people would have to meet three criteria:

1. No one who gets a piece of cake wants anyone else's piece instead. The division would, therefore, be called envy-free.
2. It would be impossible to make anyone happier than they are with their piece without making someone else unhappy. This condition is called efficiency.

3. The division would be equitable; that is, each person would believe that all the pieces had the same value. (For instance, if three people were dividing a cake and each liked frosting flowers, they might feel that a division would be equitable if each slice had a flower.)

In 2014, two researchers—Julius Barbanel of Union College and Steven Brams of New York University—published an algorithm in the *Mathematical Intelligencer* that they claim does in fact produce a division that meets all three criteria, resulting in a perfect allocation of cake. (Their method, however, assumes that there are only two people sharing the cake.) The algorithm takes into account the fact that a cake is "heterogeneous," meaning that it has different parts that two people might value differently. One person, for instance, might love the large amounts of frosting on the cake's outer edge, while the other person might like the cake itself more than the frosting. In addition, the method relies on a third party, who acts as a referee. Finally, the algorithm mentions something called a probability density function, which is just a mathematical way of representing a person's preference for different parts of the cake.

In the first step of the algorithm, each person submits his probability density function, or PDF, to the referee. (The referee could take many forms: a computer, an older sister, a stranger on the street, or a parent, among others.) The referee marks all the places on the cake where the PDFs cross; in other words, where each person's preferences overlap. The referee assigns portions based on the preferences, and if at this point each person gets equal-sized pieces, the algorithm stops and everyone eats. For example, let's say that person A likes chocolate cake and person B likes vanilla. If the cake is divided exactly equally between the two flavors, the referee can just slice the cake in half along the demarcation and give each person the half she prefers. If the cake is divided unequally though, the person with the bigger piece gives part of his piece to the other person, beginning with a region in which the ratio of his preferences is smallest. This process continues until the area of each person's portion is the same.

In addition to the Brams-Barbanel method, which helps two people divide a cake fairly, there is another, more general, method, which can help an unlimited number of people divide a cake into an unlimited number of slices. Brams and another mathematician, Alan Taylor, invented it, and it was published in the January 1995 issue of the *American Mathematical Monthly*. This general method is complicated, but the gist is that after the cake has been cut by person A, person B can trim some of the pieces to make them more equal, in case he feels that person A cut the pieces unfairly. Person C can then trim some pieces, and then person D, and so on. In addition, the method ensures that there will be extra pieces of cake left over, so if anyone feels that they have been cheated, they can always choose one of the leftovers, which will be as least as big as the piece they wanted.

NONADDITIVE UTILITY

Fair division solutions assume something called additive utility. In other words, if I like a little bit of frosting, then I'll *really* like a *lot* of frosting; the more, the better. On the other hand, if the pleasure I got from eating frosting were not additive—in other words, if it was a nonadditive utility—then after a certain amount of sugary goodness I wouldn't continue to grow any happier. Researchers have shown that fair division solutions fall apart in situations involving nonadditive utility.

42

Delivering Packages Efficiently

MATHEMATICAL CONCEPT: TRAVELING SALESMAN PROBLEM

WHEN YOU GET A PACKAGE from UPS, you might think that math has nothing to do with its getting to your door. In fact, mathematics now plays an important role in how the guys in the brown trucks deliver packages.

At the heart of UPS's operations is the process of determining the shortest route that any particular driver can take. UPS has around 96,000 delivery vehicles—ranging from cars to vans to motorcycles to alternative-fuel vehicles—and each driver visits, on average, 150 destinations every day. A driver going even one mile more than necessary on a delivery route costs the company millions of dollars per year. The incentive to make each route as short and efficient as possible is huge.

This kind of problem—finding the best possible route—is well known to mathematicians, who call it the Traveling Salesman problem. The name was coined in the days when door-to-door sales were more common; a salesman who had to visit a certain number of households in one day had to figure out the route that took him to each household in the least amount of time. Traveling Salesman problems are difficult to solve because the number of factors that must be taken into account is staggeringly high. For instance, if a driver is scheduled to visit 25 locations during his day, the number of possible routes is 15 trillion trillion. But by using computers and algorithms—sets of instructions designed for a specific purpose— UPS can whittle down the number of possible routes in a short amount of time.

UPS's efforts to perfect its algorithms ramped up in the 2000s with the creation of a computer program called ORION (On-Road Integrated Optimization and Navigation). ORION's mathematical calculations have saved UPS drivers millions of miles every year. You might do this in your own life, if you have a series of errands to run—you mentally calculate the most efficient route to make each stop, minimizing time- and energy-wasters such as retracing your steps or venturing into a busy spot during rush hour.

TRAVELLING SALESMAN: THE MOVIE

The Traveling Salesman problem has made it to the silver screen. In 2012 a movie called, appropriately enough, *Travelling Salesman* focused on four mathematicians who must decide whether to give the United States military a solution to the P=NP problem (see entry #75), knowing that the release of their work would have moral implications, since once the military had the solution it would be able to crack any code in the world, giving it unprecedented power.

How Do Algorithms Affect Your Internet Experience?

MATHEMATICAL CONCEPT: ALGORITHMS

IN ESSENCE, an algorithm is a set of instructions that tells you how to reach a definite goal in a finite number of steps. In theory, algorithms aren't limited to the realms of mathematics and computers. If you want to build a birdhouse, you have to follow a specific set of instructions. If you want to make a bowl on a potter's wheel, change a tire on your car, or make toast, you again have to follow a set of instructions. Each of those sets of instructions is an algorithm.

You're probably more familiar with algorithms than you think. In elementary school, when you learned the procedure for dividing numbers and adding fractions, you were learning an algorithm. You were also learning algorithms when you learned the order of operations (that to compute a calculation you should always begin with the stuff inside the parentheses, and then continue with exponents, multiplication, division, addition, and subtraction). In other words, whenever you calculate the tip at the end of a meal, or add numbers on the back of an envelope, you are using an algorithm.

Algorithms are particularly relevant in your everyday use of the Internet. If you are active on the web, you are interacting with algorithms constantly. For instance, whenever you order a movie that Netflix has recommended for you, you are taking advantage of the computing power of an algorithm. Whenever you search for a term using Google, customize your music preferences in Pandora by giving some songs a thumbs-up and others a thumbs-down, or browse on Amazon, algorithms make your online experience richer by communicating a sense of what you like and don't like. With that information, websites and programs can offer you specific choices based on your preferences.

THE NETFLIX PRIZE

In 2006 Netflix organized a contest to improve its recommendation algorithm by 10%. In 2009, the $1 million grand prize was awarded to a team called BellKor's Pragmatic Chaos. The key to their victory was to predict the movies that people would like based on various data, and then compare those predictions with how viewers actually reviewed the films later. And recommendations matter to Netflix: The company claims that accurate representations are part of the company's core identity.

Explaining the Monty Hall Problem

MATHEMATICAL CONCEPT: PROBABILITY

SOME EXAMPLES of mathematical thinking, like the birthday paradox (see entry #78), are odd and counterintuitive, but others are so bizarre that even professional mathematicians have a hard time believing they are true. One of those examples is the Monty Hall Problem, named after the host of the game show *Let's Make a Deal.* The solution is so surprising that even when it is thoroughly explained, most people feel that it somehow can't be right. In a way, it's the mathematical equivalent of quantum mechanics (a branch of physics that studies matter's tiniest components): strange, hard to believe, and yet correct.

On the show, the host presents the contestant with three doors. Behind one of the doors is a new car; behind each of the other two doors is a goat (or something else that isn't as cool as a car). The host asks the contestant to pick which door she thinks hides the car. Now, without opening that door, the host opens another door, showing a goat. The contestant now has the option of changing her original choice to the third door. The question is whether the contestant should stay with her original choice or switch to the other door.

The answer is that the contestant should *always* switch to the other door. At the beginning of the game, the contestant has a one in three chance of choosing the door hiding the car, but switching doors at this point doubles the odds to two in three. How can that be? Most people feel that it doesn't matter whether the contestant switches: after the host opens the door showing one of the two goats, the odds of getting the car are now one in two, since one remaining door must hide a car and the other must hide the other goat.

But that conviction is wrong. And you can see why if you get a piece of paper and write out the possibilities. The key to the problem is that the host always opens a door showing a goat. (He would never open a door showing the car; doing so would ruin the game!) Now, without relying on our intuitions, let's figure out the possible permutations:

- **Option 1:** The contestant picks the door with goat #1. The host opens the door hiding goat #2. Staying with her original choice leads to the goat; switching leads to the car.
- **Option 2:** The contestant picks the door with goat #2. The host opens the door hiding goat #1. Staying with the original choice leads to the goat; switching leads to the car.
- **Option 3:** The contestant picks the door with the car. The host opens a door hiding either goat #1 or goat #2. Staying with the original choice leads to the car; switching leads to a goat.

So you can tell from these three options that, in two out of three cases, switching leads to the car. The result is completely counterintuitive, but completely true. That's the power of mathematics.

BERTRAND'S BOX PARADOX

A similar problem is Bertrand's box paradox, named for Joseph Bertrand, who published it in a book in 1889. Imagine that there are three boxes: one with two gold coins; one with two silver coins; and one with one gold and one silver coin. Now choose a box at random, and pick out a coin (also at random). If it's gold, what are the odds that the remaining coin is also gold? You might think that the chances are one in two, but really they're two in three.

The Math Behind Juggling

MATHEMATICAL CONCEPT: COMBINATORICS

WHEN YOU THINK of juggling, you might think about birthday clowns or the circus. But you might not know that juggling has become a topic for mathematical thought, an obsession for people interested in patterns and puzzles. And, like math, juggling now has its own notation.

Called siteswapping, this notation was invented independently by jugglers at Caltech, Cambridge University, and the University of California at Santa Cruz in the 1980s. Siteswapping assigns a number to each throw. An odd number means that the object being juggled—a ball, say—goes from one hand to another, while an even number means that the object stays in the same hand. The size of the numbers is important, too: the larger the number, the higher in the air the ball is tossed. In a throw notated as 3, for instance, the ball would be thrown to about the level of the face, and in the process would be tossed to the opposite hand. In a throw notated as 2, the ball would be tossed just a few inches in the air and be caught in the same hand. Juggling aficionados can share their throwing routines by writing them down in siteswapping notation, and can use siteswaps from others to try out new patterns. Jugglers have also learned that a routine's siteswapping notation reveals how many balls are necessary for that routine. The number of balls equals the average of all the numbers in the notation, so for a 5551 routine, you would need four balls.

Juggling was also the pastime of Claude Shannon, considered to be the father of information theory. In fact, Shannon created a juggling equation: $(F+D)H=(V+D)N$. (F stands for the time a ball stays in the air, D for how long a ball remains in a hand, H for the number of hands, V for how long a hand is empty, and N for the number of balls being juggled.) Also, using parts of an Erector set, Shannon built a juggling machine that bounced small steel balls off a tight membrane. Clowns and circuses, indeed!

JUGGLING RECORDS

According to the *Guinness Book of World Records*, the UK's Alex Barron holds the record for juggling the largest number of balls. On April 3, 2012, the eighteen-year-old managed to juggle eleven balls with twenty-three consecutive catches.

46

Nash Equilibrium

MATHEMATICAL CONCEPT: GAME THEORY

MATHEMATICS DOESN'T DEAL just with the properties of numbers. Some branches of math also try to capture human behavior, especially how people interact with each other. One of these branches is game theory.

Game theory was pioneered by John Forbes Nash Jr., the mathematician at Princeton University who was the subject of *A Beautiful Mind*, a novel that was made into a movie released in 2001 starring Russell Crowe. The games studied by game theorists don't just include chess and checkers. They include any kind of human interaction in which the decisions that each person makes depend on the decisions that other people make, including business decisions, war, and all kinds of economic interactions. Game theory therefore does not just involve the bare facts or rules, but also the mental states of the players, as well as what each player *believes* about those mental states.

One central element of game theory was named for Nash himself. Called the Nash equilibrium, the term describes a game in which each player would not choose to change her strategy, even if she knew the strategies of everyone else playing the game. In other words, a game is in a Nash equilibrium if no one would benefit by changing strategies.

You may already know a famous example of a Nash equilibrium: the prisoner's dilemma. In this case, two people have been charged with a crime and are assured of getting, say, three years in jail. But the prosecutor suspects that the two prisoners are actually partners, and offers each of them a deal. (The two prisoners can't communicate with each other at all, so neither knows which decision the other person makes.) If prisoner A confesses to partnering with prisoner B, and prisoner B does not confess, A will get one year in jail—a reduced sentence—while B will get five years. The opposite is also true: if prisoner B confesses to partnering with A and A doesn't confess, then B will get one year and A will get five years. If they both confess, then each will get two years. Looking at the overall picture, it seems like the best decision would be for both prisoners to confess. But if each prisoner reasons out the best course of action, not knowing exactly what the other person is doing, they will each decide to say nothing, and get three years in jail—the original sentence—though they could have received shorter sentences by both confessing. It turns out that the case in which neither prisoner confesses constitutes a Nash equilibrium, while the other case does not.

GAME THEORY

Game theory extends into every corner of our lives, even those that seem unrelated to games and decisions. One example is a recent decision by Southwest Airlines to let people pay to board earlier and thus have a better chance of having space in the overhead compartments for bags. Everyone on the flight has the same option, so when making the decision of whether it's worth paying extra, each passenger has to take into account what everyone else might choose to do. (It turns out that paying the fee currently is the better option.)

The Math Behind a Flock of Starlings

MATHEMATICAL CONCEPT: SCALE-FREE CORRELATION

MAYBE YOU'VE SEEN a video of a large flock, or murmuration, of starlings on YouTube, or perhaps you've been fortunate enough to see one live. In any case, you were probably struck by how well the individual birds seemed to coordinate with each other, each starling flying in sync with the others around it. (No starling, for instance, ever made a right turn and collided with its neighbor.) You also might have marveled at the way in which the sudden motion of a few starlings on the edge of the flock was transmitted almost instantly throughout the group, so that the entire mass of wheeling, feathered bodies seemed to act like one organism.

This behavior follows a pattern known as scale-free correlation. When a group of individuals is organized in this way, any movement that one of them makes affects all the other members, no matter the size of the group. In a group, the speed and orientation of one starling only directly affects the speed and orientation of its seven closest neighbors, but the information quickly spreads throughout the entire flock. Their behavior adheres to a statistical pattern that is similar to the way metal becomes magnetized or how snow crystals behave before avalanches occur. (A team of scientists recently figured out that starling flocks practice scale-free correlation, by creating a computer model that reconstructed the positions and velocities, in three-dimensional space, of real starlings in flocks ranging in size from 122 to 4,268.) And the starlings seem to accomplish this feat of coordination without having a leader directing all the other starlings; instead, each starling seems to obey simple rules, like "Match the velocity of your neighbors" and "Don't collide with anyone." Despite all the analysis, though, no one knows exactly how starlings, or other animals that exhibit the same group behavior, transmit information so quickly.

ANCHOVIES

Other animals exhibit similar kinds of behavior. Anchovies, for instance, swim in large groups known as shoals, which turn and pivot much as starlings do. And anchovy shoals can be huge: one that appeared off the coast near San Diego in 2014 was estimated to contain up to 100 million fish.

48

Putting a Pile in Order

MATHEMATICAL CONCEPT: COMBINATORICS

MATHEMATICS CAN even make sense of your breakfast. Imagine that you order a stack of three pancakes at your favorite diner, and when the waiter brings them to your table not only are they all different sizes but they're stacked out of order, so that the largest one is on top, the smallest one is in the middle, and the mid-sized one is on the bottom. Let's now say that you want your pancakes in order, with the smallest on top, the mid-sized in the middle, and the largest on the bottom. Let's also say that to rearrange your pancakes you have to follow this rule: You have to take a spatula and insert it somewhere in the stack, and then flip all the pancakes above the spatula so that what was on top is now on the bottom and what was on the bottom is now on top. Using this procedure, how many flips would you need to put the stack of pancakes in order?

With the stack of three in the order described, you would need two flips. For the first flip, you would slide the spatula below the bottom-most pancake and flip the entire stack over. Now the largest pancake would be on the bottom, the smallest one in the middle, and the mid-sized one on top. At this point you could just put the spatula below the small pancake and flip it and the mid-sized pancake, so their positions were reversed. Your stack would be perfect!

But mathematicians often want to figure out rules for the general case, which in this instance would be a stack of pancakes of any number and arrangement. What is the maximum number of flips we would need to reorder a stack of n pancakes? (Mathematicians call this number Pn, meaning the pancake number.) The Pn for a stack of three pancakes is three, and that's for the hardest arrangement: smallest on top, largest in the middle, and mid-sized on the bottom. (Mathematicians often want to find the *maximum* number instead of the minimum because they want to discover an outermost boundary.)

It so happens that this is a very difficult problem. Mathematicians have found Pn when the stack has 19 pancakes—it's 22—but not for when there are more than 19. In fact, no one has found a general formula that produces the maximum number of flips needed to order a stack of n pancakes.

PANCAKE DAY

Also known as Shrove Tuesday, Pancake Day is a time for Catholics to indulge in foods made with sugar and butter before Lent, a traditional period of fasting and repentance.

49

Math Gets Its Day in Court

**MATHEMATICAL CONCEPTS: PROBABILITY AND
STATISTICS, PROSECUTOR'S FALLACY**

A LOGICAL FALLACY is an error in one's reasoning process,
so that even if you begin with facts, you end up with a wrong conclusion. Sometimes logical fallacies deal with probability, that traditional mathematical topic. And in some cases, logical fallacies that
deal with probability can help determine whether a person accused
of a crime is found guilty.

One such fallacy is the prosecutor's fallacy. When a person uses
this fallacy in an argument—in this case, "argument" doesn't refer
to a fight but to a series of reasoned propositions that are used to
establish a point—he is trying to establish the odds of a certain
event occurring. But in the process of establishing those odds he
mistakenly compares the event to an irrelevant set of occurrences.

An example will help make the inner workings of the prosecutor's fallacy more clear. A famous instance of the fallacy occurred
in the 1998 trial of Sally Clark, a British woman whose two children
died when they were just a few weeks old. The defense maintained

that both deaths were caused by SIDS (sudden infant death syndrome), while the prosecution argued that Clark murdered both children. The prosecution based its argument on the probability that any one family would have two instances of SIDS deaths. Since dying of SIDS is rare, having two instances would be even more rare. One expert witness, Sir Roy Meadows, a pediatrician, argued that the odds of two SIDS deaths in the same family were one in 73 million. But he made two mistakes:

1. The first was in thinking that the deaths wouldn't have some sort of correlation, whether genetic or environmental; instead, he made his calculations assuming that each death was completely independent of the other. But that mistake wasn't an instance of the prosecutor's fallacy.

2. The fallacy occurred when Sir Roy calculated the probability of having two SIDS deaths in the same family against the pool of instances where there were no SIDS deaths at all—that is, against the larger population whose children never developed the syndrome. In reality, that comparison is irrelevant.

The comparison that *should* have been made is the following: Of the instances in which two children in one family died at very young ages, how many of the deaths were due to SIDS, how many to murder, how many to a combination of SIDS and murder, and how many to some other cause? Later calculations by a mathematics professor at Britain's University of Salford showed that a case of double SIDS is 4.5 to 9 times more likely than a case of double homicide.

As for the trial, Sally Clark was originally found guilty, though later acquitted in 2003.

BERKSON'S FALLACY

There are other situations in which the way you look at the evidence can distort the conclusions you draw. In Berkson's fallacy, a sampling bias makes a person believe that two qualities are causally connected when they aren't; instead, the connection is due to the sampling that was taken. For example, a certain collection of short women may all be very good at speaking Spanish, but that does not necessarily mean that being short and speaking Spanish are related. The two characteristics may seem to be causally connected, but the sample might have been drawn from a location where there is a high percentage of Spanish-speaking people (an American city with a large Spanish-speaking population, for example, rather than a small town in Denmark with few Spanish-speaking residents).

50

What Does a 40% Chance of Rain Really Mean?

MATHEMATICAL CONCEPT: PROBABILITY

YOU'VE HEARD WEATHER FORECASTS all your life, but what do they mean? When the meteorologist on your local television station says that there is a 40% chance of rain tomorrow, what exactly is he saying?

Predicting the weather involves probability, a fundamental branch of the mathematical family tree. And the part of the weather forecast that predicts precipitation is appropriately called the PoP, or probability of precipitation. But often people don't understand what the 40% means. It doesn't mean that rain (or snow or hail or sleet) will fall 40% of the time, or that it will fall over 40% of a given territory. Instead, it means that on ten days with approximately the same conditions that tomorrow will have, precipitation will fall on four of those days. (Conversely, the forecast also entails that precipitation will *not* fall on six of those days.)

The meaning of the forecast can be made even more precise. According to the National Weather Service (NWS)—the federal agency in charge of providing weather-related data to the nation—stating that there is a 40% chance of rain means that there is a four in ten chance that 0.01 inch—1/100th of an inch—of precipitation will fall somewhere within the forecast area. The NWS uses a specific formula to calculate the probability of precipitation: $PoP = C \times A$. In this equation, C is the confidence that precipitation will fall somewhere in the forecast area, and A stands for the percentage of that area on which that precipitation will fall.

But methods used to calculate the probability of precipitation vary. At the Weather Channel, for instance, a 40% chance of rain means that there is a four in ten chance that some precipitation, not just 0.01 inch, will fall in the forecast area, and that's including three hours before or after the forecast time. This method ensures that viewers will err on the side of caution when deciding whether to pack their umbrella.

ENSEMBLE FORECASTING

Another method is ensemble forecasting, which uses multiple predictions, each of which is likely and each of which begins with slightly different conditions. By observing how much each of the forecasts differ, meteorologists can determine how much uncertainty the future weather holds: the more the initial forecasts diverge, the less sure meteorologists are about the course of a weather event, like a storm. For instance, while Hurricane Katrina was churning in the Atlantic Ocean, and before it had moved into the Gulf of Mexico, the forecasts for where it would make landfall, and the path it would take over the continental United States once it did, varied widely. Some predicted that Katrina would hit New Orleans, while others predicted it would keep to the eastern section of the Gulf. Once Katrina crossed Florida, though, the predictions started to converge and meteorologists had a much better sense of where the hurricane would go.

51

Math-Based Test-Taking Strategies

MATHEMATICAL CONCEPT: ARITHMETIC

THE NEXT TIME you take a test, consider using math to improve your score. In an interesting twist, you can improve your score even if the test isn't about math!

In *How Math Can Save Your Life*, mathematician James D. Stein outlines a strategy that, he claims, can raise your score by one letter grade (assuming, of course, that you've gone to class and studied the material). The first part of the strategy is learning how the test will be scored. Will each question be worth the same number of points? And will you be penalized for answering a question incorrectly, as you are when you take the SAT? If you are not penalized, then you should do your best to answer every question, even if you choose the answer randomly. (Stein's strategy works only for certain kinds of tests: true/false, multiple choice, and those with problem-solving questions, like the ones found on math and science tests.)

Stein then recommends that you make a first pass through the test, answering only those questions whose answers you know off the top of your head, as well as those whose answers you know how to find. If you need to take more than one or two seconds, you should stop and move on to the next question. Then, you should count how many problems remain in the test and figure out how much time you have left. Dividing one by the other will give you the average amount of time you should spend on each remaining question.

In addition, you should definitely not try to answer the hardest questions first. You risk spending most of your time on a few questions when you could be racking up points by answering a larger number of easier questions.

MULTIPLE CHOICE

You might have been told that when you're taking a multiple choice test and don't know an answer, you should always choose "C." That might not be the best strategy, especially since your teachers are most likely aware of it and have designed their tests so the correct answers are evenly distributed. It's better to narrow down the choices and make an educated guess. But if there are four choices, you have a 25% chance of getting the answer correct with a random shot in the dark.

52

Your Immune System Can Do Math?!

MATHEMATICAL CONCEPT: TRAVELING SALESMAN PROBLEM

WE DON'T NORMALLY think of the cells in our bodies as being able to solve math problems, but recent research indicates that these tiny living things can do just that. The cells in question are a certain kind of white blood cell in the body's immune system, whose job is to locate and gobble up intruder cells (like viruses or bacteria). One challenge for these cells, once an infection has been detected, is how to most efficiently attack the invaders. This is actually a version of the Traveling Salesman problem (see entry #42). In the case of the human body, the problem is a little more urgent than that of selling to the most households: The more efficiently the white blood cells can find and destroy the intruders, the less chance that the body will be damaged.

The method the cells use is called chemotaxis, meaning that the cells find the intruders by, in a way, smelling them. They detect the invaders' chemical signatures and move toward them. According to simulations performed by computers, when white blood cells are faced with ten viruses or bacteria, the order in which they attack and eliminate the intruders is only 12% longer than the shortest possible route. Pretty impressive for tiny living creatures smaller than the head of a pin!

ARTIFICIAL IMMUNE SYSTEM

The immune system has inspired computer scientists to create a new field: artificial immune systems (AIS). A major effort within AIS is to figure out how to incorporate a natural occurrence like an immune system's memory into solving problems in mathematics and engineering. More broadly, AIS fits within the field of artificial intelligence, and is offering inspiration for new sources of ideas and innovation.

53

How Google Translate Works

MATHEMATICAL CONCEPTS: PROBABILITY, COMPUTER PROGRAMMING

IF YOU'VE EVER learned another language, then you're probably familiar with the process of translation. Armed with a dictionary and knowledge of grammatical rules, the language student painstakingly analyzes each sentence, figuring out what each word means. She then determines gender and number and picks out context clues. Unless you are fluent in both languages in question, the process is laborious and piecemeal.

But Google Translate bypasses all of this work. Google's language translation program instead uses statistics to compare documents written in the first language with documents written in the second language. Relying on text supplied by the United Nations, which regularly publishes in six languages (English, French, Russian, Spanish, Chinese, and Arabic), the program has built up a huge database of language examples. (Google Translate's database now includes information in about eighty languages.) It scans hundreds of millions of documents looking for patterns, trying to figure out how words are most often translated. This process, which doesn't rely on knowing definitions or grammar at all, is called statistical machine translation. The connection to math is that this kind of translation relies on probabilities: given a sentence in language A, what are the odds that a sentence in language B is that first sentence's translation?

Statistical machine translation has roots in information theory, a kind of applied mathematics that deals with signal processing, data compression, and languages, and is commonly thought to have been born with the 1948 publication of a paper, "A Mathematical Theory of Communication," by engineer and mathematician Claude Shannon, in the *Bell System Technical Journal*. Information theory is used in code breaking, as well as in transmitting messages using cell phones and computers. Without the underlying mathematics of information theory, the phone in your pocket would be a brick. And the amazing ability to translate text using web-based computing would be impossible.

SEISMIC OIL EXPLORATION

Information theory is also important to people who look under the ground for oil. Their field, seismic oil exploration, relies on information theory to weed out unwanted noise—data that might interfere with the signals produced by oil deposits—and produce a clear signal.

54

Don't Tailgate

MATHEMATICAL CONCEPT: ARITHMETIC

THE FASTER YOU DRIVE in your car, the more you risk getting injured. At high speeds, the time you have to react to other cars on the road goes down, while the severity of any resulting crash goes up. But by how much, exactly, does each quantity go up or down? Math can provide some definitive answers—and possibly inspire you to drive more safely.

Let's say you're driving at sixty miles per hour. If you know that a mile comprises 5,280 feet, you can calculate that you are moving at 88 feet per second. And because a car is around 15 feet long, you would be traveling the distance of 6 car lengths in that second (since 6 × 15 = 90, which is close to 88). If, as many people learned in driver's ed, you should stay one car length behind the car in front of you for every ten miles per hour you add to your velocity, you will be six car lengths behind the car in front of you (assuming that it, too, is traveling at sixty miles per hour). These calculations show that if the car in front of you suddenly got a flat tire, you would have *one second* to react.

That's why tailgating is clearly not a good idea.

THE GADD SEVERITY INDEX

The Gadd Severity Index quantifies how much a car crash affects human beings inside a vehicle. The equation looks like this: $GSI = a^{5/2}(t)$, where a is the acceleration and t is time in seconds. The human head can withstand GSI values of up to 1,000, as long as the time duration is very short (on the order of milliseconds).

55

Brazil Nut Effect

MATHEMATICAL CONCEPT: GRANULAR CONVECTION

IT'S INEVITABLE. Whenever you buy a can of nuts, the large ones—as if by magic—are all at the top. The same thing happens with breakfast cereal: the large components, like nut clusters, rise to the top, leaving the middle and bottom of the cereal box perilously empty of those tasty morsels. Besides being frustrating, the so-called Brazil nut effect has links to mathematics. But how?

One popular hypothesis traces the origin of the effect to the sizes of the particles (which could be nuts, cereal flakes, pebbles, marbles, or any other object that might be in a mixture). Whenever a mixture of particles is jostled, the particles move vertically, if only for a short distance. At that point, space opens up between the particles, and other particles on the side of the container move in to fill them. But large particles can't fit into the spaces vacated by the small particles. As a result, the large particles continually move up toward the top of the mixture. Once they reach the top, they remain there, while the smaller particles move to the sides and then down to the bottom in a continual cycle known as granular convection. (You have seen convection in action if you've watched a pot of boiling water. The molecules of water rise to the top as they get hotter, and then fall as they cool down.) Math in a box of cereal? You bet.

BRAZIL NUTS AND AVALANCHES

People who trek in icy mountains can now wear devices that inflate in case of an avalanche, making them larger and therefore more likely to move to the surface if submerged under a river of snow. The idea utilizes the same principles of the Brazil nut effect to potentially save lives.

56

Myth Busted: More Roads Don't Mean Less Traffic

MATHEMATICAL CONCEPTS: NETWORKS AND SYSTEMS, BRAESS'S PARADOX

IN 1968, German mathematician Dietrich Braess discovered an odd feature of networked systems, one that seemed to defy common sense. Dr. Braess, who now teaches math at Ruhr-Universität in Bochum, Germany, was studying traffic flow when he noticed that in some instances when the flow of cars was congested, adding roads actually made the congestion worse. Similarly, removing roads from some congested traffic areas made the cars move more easily. This notion was not only counterintuitive; it also contradicted the tenets of city planning. How could it be?

At the core of Braess's discovery is the notion that drivers are selfish. They don't coordinate their driving plans with other drivers, and everyone wants to take the quickest route from point A to point B. For instance, imagine that there are two ways to drive from the center of town to a suburban mall. Each path consists of two parts:

one section of road that drivers can always traverse in thirty minutes, and another section of road that is narrower, so that the time it takes to traverse this section depends on the number of cars on it. (We can say that the time it takes to travel along this second section is $T/5$, where T = the number of cars on the section of road.) We can also say that while both paths from the town center to the mall include both sections of road, they occur in opposite orders. (That is, on path A, the narrow road comes before the thirty-minute road, and vice versa for path B.)

How long would it take 200 drivers to get from downtown to the mall? Well, since both paths are the same—the only difference is that the sections are flip-flopped—we can assume that half of the drivers would take one and half would take the other, and thus the travel time on each path would be fifty minutes.

A driver on one of these paths would have no reason to try to use the other path, since he wouldn't save any time. (In this kind of situation, when there are many individuals, and each individual has a sense of what the others would do in his situation, and no one has an incentive to change his strategy, the individuals are said to be in a Nash equilibrium—see entry #46.)

Now let's say that the town constructs a shortcut between the two paths, at the point along each path where the two sections meet. This shortcut takes almost no time to traverse. The drivers now might reasonably want to use the same route: they could take the $T/5$ section of path A, then the shortcut, and then the $T/5$ section of path B. (This path would form a kind of zig-zag shape.) But of course *all* the 200 drivers would want to take this route, in an attempt to decrease travel time, meaning that the trip would take $200/5 + 200/5$, or eighty minutes. The drivers will all know that the shortcut could possibly reduce their driving time, and thus they all will try to take that route. As a result, the traffic flow worsens.

The idea that reducing the amount of choice can produce better driving conditions has been used in real-life cities, including Seoul, the capital of South Korea. When a six-lane highway running through the center of the city was torn down in the mid-2000s and replaced by a five-mile-long park, traffic patterns actually became more efficient. Cars were diverted onto roads that already existed. The result seemed to defy common sense, but math helped reveal its wisdom.

POWER LINES

Braess's paradox doesn't just apply to traffic. In a 2012 paper, scientists at the Max Planck Institute for Dynamics and Self-Organization showed that adding power lines to an energy grid doesn't necessarily improve the grid's performance. Instead, the new lines can end up destabilizing it, depending on where they are located in relation to existing lines; thus, fewer lines sometimes result in a more efficient power grid.

57

How Many Times Can You Fold a Sheet of Paper?

MATHEMATICAL CONCEPTS: EXPONENTIAL GROWTH, BEDSHEET PROBLEM

HOLD A PIECE OF PAPER in your hands. Fold it in half. Now fold it in half again. How long do you think you can continue folding? This is a mathematical puzzle known as the bedsheet problem, but it could just as easily apply to paper, towels, aluminum foil, noodles, and anything else that one can fold. For years, mathematicians thought that the maximum number of folds was seven. In 2002, though, Britney Gallivan, a high-school student in Pomona, California, set a record by folding a very long sheet of toilet paper—4,000 feet long, to be exact—in half twelve times. She did so by only folding in one direction, and only after working out the calculations that prescribed how long a sheet of paper—or anything—would have to be.

So what? Well, folding a sheet of something in half over and over again is a good way to understand exponential growth. When a size or number grows exponentially, it increases by a set exponent at each step, but because the base number increases each time, the resulting number gets large very quickly. For instance, let's consider a sheet of regular loose-leaf paper, which is probably around one-tenth of a millimeter thick. After the first fold, the paper will be two-tenths of a millimeter; after the second fold, the thickness will be four-tenths of a millimeter. After 25 folds, the piece of paper would be 1 kilometer thick. After 42 folds, it would be thick enough to reach the moon. After 81 folds, the paper's thickness would span 127,786 light years. And after 103 folds, the paper would take up more space than the visible universe (a distance of about 93 billion light years).

THE TOILET PAPER PROBLEM

Computer scientist Donald Knuth once wrote a paper on two-roll toilet paper dispensers in public buildings, in the process classifying people into two groups. Big-choosers first reach for toilet paper from the larger roll; little-choosers first use the roll with the lesser amount. His paper studied the probability of someone being one or the other, and how that impacts the amount of sheets left on the rolls, using different mathematical equations.

58

Yes, There *Is* a Better Way to Board an Airplane

MATHEMATICAL CONCEPT: EFFICIENCY

BEING ABLE TO FLY from Los Angeles to New York City in five hours is a miracle, but having to endure the boarding process makes that miracle a chore. Passengers usually board planes in blocks, from the back to the front, and while this method is meant to prevent congestion, there are inevitable delays as people take time to shove their carry-on bags into the overhead compartments. In addition, people who have window seats often have to wait for people in middle and aisle seats to get up before they can sit down. All of these factors create a headache for the weary traveler, and the wasted time costs airlines money.

Mathematicians have put their brains to the challenge of making boarding planes less of an ordeal, though, and they have come up with a solution. The secret starts with spacing and seat type. People in the odd-numbered rows should board first. This way, there is always an empty row between people stowing their baggage, giving them room to maneuver. An added requirement is that of those people in odd-numbered rows, those who have window seats board before anyone else. Then come people in odd-numbered middle seats, and finally people in odd-numbered aisle seats. This method ensures that no one has to climb over anyone else, and the time to get to one's seat is minimized. The entire process repeats for the even rows. Simulations show that this method is so efficient that passengers might be able to board in one-sixth the time of the current block-by-block method. So why aren't airlines using this math-approved technique? Maybe mathematicians can get to work answering that question next.

SOUTHWEST

Southwest Airlines does not assign seats, and thus people are free to choose however they like, according to the number on their boarding pass. (This number is assigned when fliers check in, but for a fee they can get a better boarding number—see entry #46.) It's unclear whether unassigned seating is actually a more efficient method, since randomness is invited into the equation.

PART 3
PATTERNS

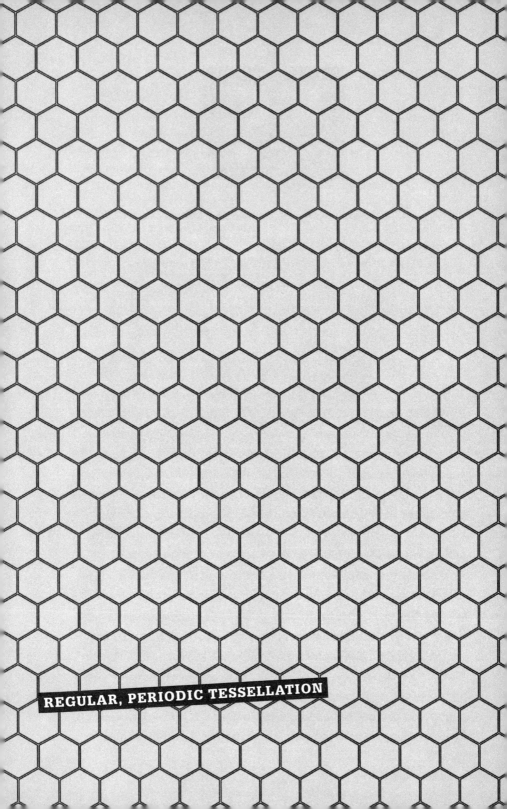

REGULAR, PERIODIC TESSELLATION

59

Tessellations

MATHEMATICAL CONCEPT: GEOMETRY

THAT M.C. ESCHER POSTER that hung on your dorm room wall has more of a connection to mathematics than you might think. Escher's drawings are examples of tessellations, the covering of a two-dimensional plane, such as a sheet of paper, with geometric shapes so that the shapes don't overlap and there aren't any gaps between them. As Escher's artwork proves, those shapes don't have to be triangles and squares; they can also be birds, angels, fish, or blobs. In fact, you can think of a jigsaw puzzle as a kind of tessellation. The pieces fit together to completely fill the plane of the completed puzzle, with no overlaps and no gaps. But tessellations aren't just found in Escher posters. From the intricate tile work of Spain's Alhambra, to the six-sided cells in a bee's honeycomb, to the geometric patterns that covered walls and floors in ancient Roman buildings, to shapes on quilts, tessellations are everywhere.

Tessellations have also proven to be a fertile realm of mathematics. Over the centuries, mathematicians have found that tessellations occur in particular forms:

- Some tessellations are periodic, in which patterns repeat, and some are nonperiodic, in which they don't.
- Some tessellations are regular: they're formed by repeating one regular polygon, a shape whose lines are all the same length and angles are all of the same size. (Think of a square.)
- Other tessellations are semiregular, meaning that they are made up of more than one regular polygon.

The analysis goes further. In 1891, a Russian crystallographer named Evgraf Fedorov proved that periodic tessellations fall into one of seventeen categories. And there are eight kinds of semiregular tessellations.

All of which goes to show that mathematics is not just about calculating. It can also be about the wonder and beauty of shapes.

MAURITS CORNELIS ESCHER

Born in the Netherlands, graphic artist Maurits Cornelis Escher failed the exams that would have let him study architecture. But a later visit to the Alhambra, a fourteenth-century Moorish palace in Spain, inspired him to concentrate on creating designs that completely filled the plane. The rest is history.

60

There Are 177,147 Ways to Tie a Necktie

MATHEMATICAL CONCEPTS: GEOMETRY, TOPOLOGY

THE INSPIRATION for new mathematics can come from anywhere. For instance, one mathematician—Mikael Vejdemo-Johansson at the Jožef Stefan Institute in Slovenia—got an idea while watching one of the Matrix movies. He noticed that a character known as the Merovingian sported a series of ties with unusual knots; one of them stood out because it looked like the tie itself was wearing a tie. Intrigued, Mikael did some research and found that a team at Cambridge had published a paper about the mathematics of knotting ties. The two researchers had broken down the act of tying a tie into a series of steps that could be represented by symbols. (Some of these include "L" for left, "C" for center, "R" for right, "i" for a move that brings the active piece of tie into the diagram, "o" for a move out of the diagram, and "T" for a move that brings the thread through the bow of the knot.) They divided the possible tie knots into 101 classes—depending on the overall number of moves, as well as the number of centering moves—for a total of 85 knots.

The problem, however, was that the Merovingian's knot was nowhere on that list. Vejdemo-Johansson discovered that the Cambridge researchers—Yong Mao and Thomas Fink—had made two assumptions that ended up limiting the number of possible knots. First, they stipulated that all the knots would be covered by a flat piece of the tie's fabric, and second, that a tucking move would only occur at the end of the knot-tying process. Vejdemo-Johansson found that if he simplified the formal language of representing tie knots and increased the number of times that one end of the tie could be wrapped around the other from 8 to 11, he would end up with a total of 177,147 possible knots. He and his two teammates found 2,046 categories for winding patterns, which can require up to 11 moves to complete.

So the next time you find yourself bored with the current way you tie your tie, remember that you have enough options to last you for the rest of your life!

NECKTIE KNOTS

One of the easiest necktie knots is the four-in-hand, which is probably what you learned when you first began wearing ties. A more complicated knot is the Windsor, made popular by the Duke of Windsor and best suited for shirts with wide collars.

61

Little-Known Connections Between Music and Math

MATHEMATICAL CONCEPTS: NUMBER THEORY, RATIOS

MUSIC AND MATH have always had a tight relationship. From the time of the Pythagoreans and the ancient Greeks to the compositions of Bach, which sometimes sound like theorems converted into sound, to the intricate workings of musical notation—full of quarter notes, scales, and tempos—music has embodied mathematics in a way that not many other disciplines have. In one sense, the mathematics of music is obvious: numbers are found throughout. For instance, some pieces are written in 4/4 time, while waltzes are in 3/4 time, and some Slavic music is written in 12/16 time. Some notes are held for the duration of a measure, others for only $\frac{1}{16}$ of that measure. Tempos are referred to in beats per minute. Meters tell the musician how many beats occur in each measure and which kind of note should receive one beat. No matter where you look, music is permeated by math.

In another sense, though, the math of music is not obvious, but this hidden math is the backbone of all music, no matter where in the world it occurs. This occult mathematical aspect is the characterization of musical intervals. Play any two notes simultaneously on a piano and the resulting note combination will sound either pleasant or ugly, full or thin.

One of the most pleasing two-note combinations, or intervals, is the octave, which is the relationship between one note and another at either double or half the original note's frequency. If you were looking at a piano keyboard, an example of an octave would be middle C played with either the next highest or next lowest C. (The two Cs would be eight white keys apart.) Octaves can also be expressed as ratios. Because one note in every octave has a frequency twice as high as the other note's frequency, the ratio would be 2:1. Other intervals have their own ratios, as well as adjectives like "perfect," "augmented," and "diminished." (The term "perfect" usually refers to intervals that sound especially pleasing to most human beings. Augmented refers to a perfect interval that has an added semitone, or half-step. For instance, while the notes C and G played together form a perfect fifth, the notes C and G-sharp—one black key, or semitone, up from G—form an augmented fifth.) The ratio for the perfect fifth is 3:2, while that for the major third—an interval consisting of four half-steps—is 5:4. Thinking about combinations of notes in terms of ratios helps reveal the underlying math of the music we hear every day.

UGLY MUSIC

Using techniques developed in the 1950s to improve naval sonar, mathematician Scott Rickard created a completely patternless, though not random, piece of music that he calls the "world's ugliest."

62

The Game of Go

MATHEMATICAL CONCEPT: COMBINATORICS

MANY GAMES have mathematical underpinnings, but perhaps none are as elegant as the game of Go. Thought to have been invented in China approximately 4,000 years ago, Go is hugely popular in China, Japan, and Korea and has been slowly making its way into the Western consciousness. (For instance, the American Go Association was only formed in 1935.) The rules are simple: One player has a collection of black stones, while the other has a collection of white stones. The board, often made of wood, is divided into a 19 × 19 grid—that is, a grid formed of nineteen rows and nineteen columns. Players place stones at the intersections of the grid lines with the goal of acquiring and defending territory. You can capture one of the other player's stones by surrounding it with your stones. Once surrounded, the captured piece is removed from the board.

Go practically oozes with mathematics. For instance, consider the number of legal positions: a little more than 2×10^{170}, which sounds even more enormous when you learn that the number of atoms in the known universe is only around 10^{84}. Big numbers also appear when you compare Go to chess. When a computer program plays chess, it can analyze the consequences of each move up to seven turns in advance. But if a computer were to apply this technique to Go, it would quickly become overloaded. When playing chess, a computer might sift through up to 60 billion possibilities at each turn. Looking ahead seven moves in Go, however, would require a computer to sift through 10 thousand trillion possibilities.

The game also helped give rise to an entirely new class of numbers. In 1970, mathematician John Conway, at Cambridge University, was studying a game of Go being played by two masters and eventually came up with the idea of surreal numbers. You can think of surreal numbers as sets of instructions to find particular numbers on the number line using a series of up and down movements. All the real numbers—which consist of whole numbers, fractions, positives, negatives, and irrationals—count as surreals, but some surreal numbers are not real numbers. The surreals are in essence a new set of numbers, like the rationals and the integers, that you find on the number line using a series of moves—ups and downs, or lefts and rights. One particularly large surreal number is known as omega, and it's defined as the number on the number line you land on when you go to the right for an infinite amount of time. (Omega is the smallest surreal number that's larger than any of the real numbers.) In any case, Go was the impetus for the discovery, and the game remains a mathematical treat for millions around the world.

OTHELLO

A game that resembles Go is Othello, invented in the 1880s by two Englishmen and originally known as Reversi, though the games are actually quite different. While players in both games surround the pieces of their adversaries, in Othello the surrounded pieces, which are black on one side and white on the other, are flipped over. In Go, captured pieces stay the same color. In addition, the Othello board, with a grid of 8 × 8, is far smaller than the 19 × 19 Go board.

63

Chessboard and Wheat

MATHEMATICAL CONCEPT: GEOMETRIC PROGRESSION

THERE IS A STORY that Sissa Ben Dahir, a vizier in the court of King Shirham of India, invented the game of chess. Pleased with Dahir's invention, King Shirham offered to give him anything he asked for. Sissa Ben Dahir asked for a seemingly innocuous gift: one grain of wheat for the first square of the chessboard, two for the second, four for the third, and so on, with each square receiving twice the number of wheat grains than the square before it. We can represent that operation as a series of additions: $2^0 + 2^1 + 2^2 + 2^3 + \ldots 2^{63}$. (We stop at 63 because although a chessboard has 64 squares, the exponent of the first 2 in the series is 0, not 1.) This kind of addition, in which the base stays the same while the exponent increases with each step of the progression, is known as a geometric progression. And while the sum may not seem like it would be large, it definitely is. In fact, it's the same number as the number of moves required to solve the 64-disk Tower of Hanoi problem: 18,446,744,073,709,551,615 (see entry #64). Assuming that there are around 100 million grains in a ton of wheat, Sissa

Ben Dahir's request equals approximately 200 billion tons of wheat, a truly astonishing amount.

LEWIS CHESSMEN

One of the coolest chess collections in the world is known as the Lewis chessmen. It consists of ninety-three pieces dating from the twelfth century, found in the Outer Hebrides of Scotland in 1831. Made of walrus ivory and whale teeth, they seem to have Nordic influences: the rooks are shaped like soldiers biting their shields, like Viking berserkers.

64

The Tower of Hanoi

MATHEMATICAL CONCEPTS: RECURSION, GEOMETRIC PROGRESSION

SOMETIMES SIMPLE RULES can lead to astoundingly big numbers. Consider the Tower of Hanoi, a toy consisting of three sticks set upright in a hard base, and a stack of wooden disks—each with a hole in its middle—speared by one of the sticks. Each disk has a different size, and they are stacked in such a way that the smallest disk is at the top, the next-largest below it, and so on until we reach the largest disk at the bottom. The goal of the game is to move the stack from one stick to another, with the disks ending up in the same order, but rules stipulate that you can only move one disk at a time, and you can never put a disk on top of a smaller disk.

The moves required to achieve the goal are an example of recursion. Moving the first disk requires one move, but each subsequent disk requires double the number of moves of the previous disk. With enough disks, the number of moves to finish the puzzle is inconceivably large. For instance, there is a legend about the Tower of Hanoi related in *Mathematical Recreations and Essays*, a book by W.W. Rouse Ball and edited by H.S.M. Coxeter. According to the legend, there exists in India a Tower of Hanoi with three diamond needles, and on one is stacked 64 golden disks, each one smaller than the one below it. Priests of Brahma tend the disks, and at all times one of the priests is in the process of moving the disks to another needle, according to the simple rule just mentioned. When the entire stack has been moved, the world will end.

How long will this task take? If each move takes one second, and the priests never take a break, moving the stack of golden disks will take 18,446,744,073,709,551,615 seconds, which equals about 58 trillion years, far longer than the current age of the universe (which is only 13 billion years or so). Huge numbers truly can be contained in simple packages.

TOWER OF HANOI IN POP CULTURE

The Tower of Hanoi is a popular device in pop culture. In a 1966 episode of *Doctor Who*, the Celestial Toymaker forced the Doctor to play a ten-piece version of the game, with a limit of 1,023 moves, which he called the Trilogic game. In the 2011 movie *Rise of the Planet of the Apes* this puzzle, here called the Lucas Tower, was used to test the intelligence of apes.

65

Pigeonhole Principle

MATHEMATICAL CONCEPTS: PIGEONHOLE PRINCIPLE, COMBINATORICS

NEVER DISCOUNT A SIMPLE IDEA, because those ideas sometimes have wide-ranging implications. One such idea is the pigeonhole principle, first formulated in 1834 by German mathematician Peter Gustav Lejeune Dirichlet. According to this principle, if you have three pigeonholes and four pigeons, and all the pigeons have to go into a pigeonhole, then one pigeonhole *must* contain more than one pigeon. (The principle doesn't tell us exactly how many pigeons are in each hole, or even that each hole has a pigeon. For all we know, all four pigeons decided to share the same pigeonhole, leaving two holes empty.) If we want to write the principle in a more general way, without referring specifically to pigeons (the principle works just as well with cows, turkeys, footballs, or any other object), we can say that if we have N sets and M objects, and M is greater than N, then one of the sets will contain at least one object.

You can use the pigeonhole principle to make claims about the world. For example, let's say you have a bag of M&M's, half of which are red and half of which are brown. What is the minimum number of M&M's you would have to take from your bag to make sure you have at least two of the same color? (The answer is three. You might choose two of the same color at the beginning, but you might also choose one that's red and one that's brown. In that case, no matter which M&M you choose next—the third—you will have a pair. In this scenario, you can think of the two pigeonholes being two boxes: one for red M&M's and one for brown. We want to find the fewest number of M&M's we must take from the bag to ensure that two of them end up in the same box.)

You could also use the principle to determine that two people in New York City must have the exact same number of hairs on their heads. Each person's head has around 100,000 hairs, and there are around 8 million people living in New York. Since there are 100,000 possibilities for the number of hairs on any particular person's head, there are 100,000 pigeonholes, so to speak. And the 8 million New York City residents correspond to 8 million pigeons. So we have 100,000 pigeonholes and 8 million pigeons: therefore we can be certain that at least two pigeons—or people—occupy the same box, meaning that the number of hairs on their two heads is probably the same.

PIGEONHOLING IN CONGRESS

Pigeonholes also appear in contexts that have nothing to do with either pigeons or mathematics. In Congress, to pigeonhole a bill is to place it aside, as if in a cubby, and ignore it for the moment.

66

Mazes

MATHEMATICAL CONCEPTS: GRAPH THEORY, TOPOLOGY

MAZES HAVE LONG BEEN a part of popular culture, from the myth of Theseus and the Minotaur; to meditational church mazes constructed during the Middle Ages; to maize mazes, which pop up in rural communities during autumn; to movies like *Labyrinth* and *The Maze Runner*. But while they are intriguing and beautiful in their own right, they also are part of the family of mathematical objects.

The study of mazes belongs to graph theory and topology, fields that examine objects in a schematical way (much like the analysis we saw in entry #9 on subway systems). If you think of a maze in an abstract way, ignoring the twists and turns you might have to take, or the height of the walls, or the texture of the ground beneath your feet, you can think of it as a path that, at certain points, forks into new directions. We can call each of these points a node. A path connecting two nodes next to each other is called an edge. If we were looking at a maze from above, we could take notes and then make a drawing, a kind of diagram that consisted only of nodes and edges. After labeling the nodes, we could see more easily the path we would have to take to get to the maze's end.

This kind of analysis was first conducted by Leonhard Euler, a Swiss mathematician who lived in the 1700s. He tackled and solved a problem known as the Seven Bridges of Königsberg, and, in doing so, founded the field of graph theory. The problem was based on the real-life city of Königsberg, in Prussia. The Pregel River ran through the city, and in the middle of the river was an island. After the river passed the island, it forked into two branches. Seven bridges connected the island to the mainland, and locals wondered whether there was a way to cross each bridge only once and return to one's starting point. By thinking of the bridges, island, and mainland as an abstract network consisting of nodes and edges, Euler was able to prove that such a path did not exist.

THE MINOTAUR

One kind of maze, known as a labyrinth, has only one path that leads directly from the entrance to the center. One famous labyrinth is said to have been constructed by King Minos below the palace of Knossos around 3,000 years ago on the island of Crete. According to legend, King Minos built the labyrinth to hold the Minotaur, a creature born of the union between the queen and a bull. Minos had forced the people of Athens to send him seven young men and women each year, to be placed inside the labyrinth for the minotaur to eat. Theseus decided to end this horrific tradition. He volunteered to be one of the tributes, and when they were all presented to the king, the king's daughter, Ariadne, fell in love with Theseus. She gave him a ball of string to unwind as he navigated the labyrinth so he could find his way back out. Theseus slew the Minotaur and escaped the labyrinth, but on the way back the Athens forgot to change his ship's sails from black to white, a signal prearranged with his father to show that he had survived the Minotaur encounter. Theseus's father, Aegeus, saw the sails, and despondent with grief, threw himself into the ocean.

67

How Many Clues Do You Need to Solve a Sudoku Puzzle?

MATHEMATICAL CONCEPT: NUMBER PUZZLES

SUDOKU MAY BE one of the nation's favorite pastimes, but it's not just a way to spend a few spare minutes (or hours). The addictive number puzzle also contains some interesting mathematical nuggets within its beguiling borders.

Sudoku consists of a 9 × 9 grid, with each square further divided into a smaller 3 × 3 sub-grid (think of an enormous waffle iron). Within each sub-grid, players must fill in the numbers 1 through 9 so that each number appears only once in each row and column of the larger 9 × 9 grid. In addition, each number must appear only once in each 3 × 3 sub-grid. Scattered throughout the grid are numbers that have already been entered by the puzzle setter (see image); these are clues that guide how the player solves the puzzle. Another feature of sudoku is that each puzzle has one, and only one, solution.

A team of mathematicians, led by Gary McGuire at University College Dublin, has discovered that the minimum number of clues a puzzle must give to have a unique—that is, only one—solution is seventeen. If there are fewer clues provided, the puzzle cannot have a unique solution. McGuire and his team did not find a proof, though. Instead, they used brute computing power to search through all possible sudoku grids. In fact, they used about 7 million hours of computing time at Dublin's Irish Centre for High-End Computing. They needed all the computer power they could get, because the number of possible sudoku grids is huge: 6,670,903,752,021,072,936,960. The researchers were able, however, to cut that number down to a more manageable size using an algorithm based on the principle that some of the different grids are mathematically equivalent.

All this goes to show that even a diversion in your local newspaper can be full of interesting mathematics.

				4	9		6	
5		6	3	9			1	
				6				
8			4			3		5
	6		5		2		4	
3		4			8			2
			6					
	4			2	3	8		1
1		3	8					

NP-COMPLETE

In 2002 mathematicians asserted that sudoku is NP-complete. ("NP" stands for nondeterministic polynomial time). What does that mean? In essence, there is no fast, easy way to solve a sudoku puzzle, even though it's very easy to confirm whether a given solution is correct. NP time is very long. What does that mean for sudoku? In essence, there is no fast, easy way to solve a sudoku puzzle, even though it's very easy to confirm whether a given solution is correct.

68

Mathematical Patterns in Van Gogh's Work

MATHEMATICAL CONCEPT: TURBULENCE

THE STARRY NIGHT is one of Vincent Van Gogh's most beautiful and iconic paintings, but recently it has become known not just for its beauty but also for its underlying mathematics.

It turns out that the swirling patterns of *The Starry Night*, as well as those in *Wheatfield with Crows* and *Road with Cypress and Star* (two other Van Gogh paintings), display an uncanny resemblance to turbulent flows. Turbulence is the kind of motion you might find in river eddies or in smoke rising from a fire. Turbulence also occurs in the motion of fluids flowing through pipes, and the turbulent mixing of warm and cool air in the atmosphere is responsible for the jerky motion we sometimes feel while in an airplane. Though turbulence is common, describing it using mathematics is famously difficult. To do so, mathematicians would need to understand the solutions to the Navier-Stokes equations, formulated in the 1800s, which describe the motion of fluids. Those equations are in fact very hard to solve. (There is a story involving turbulence and physicist Werner Heisenberg. When asked what he would say to God, if he had a chance, Heisenberg said, "When I meet God I am going to ask him two questions: Why relativity? And why turbulence? I really believe he will have an answer for the first.")

To determine whether the patterns in *The Starry Night* matched the characteristics of turbulent flow, scientists examined the luminance, or brightness, of the paint laid down by Van Gogh's brushstrokes. They studied a digital version of the painting and compared the luminance of the pixels within the image. They found that the patterns of luminance matched equations formulated in the 1940s by Russian mathematician Andrei Kolmogorov, when he was trying to understand turbulence using statistics. The swirls of Van Gogh's brushwork truly have deeper meaning.

ANDREI KOLMOGOROV

Andrei Kolmogorov was born in 1903, the son of an agricultural researcher. Kolmogorov had wide interests: in mathematics he studied probability, topology, and turbulence, among other topics. He also devoted himself to studying history and reforming education in the Soviet Union. He died in 1987.

69

Why It's a Mathematical Feat for You to Walk Across a Room

MATHEMATICAL CONCEPTS: ZENO'S PARADOX, INFINITY, INFINITE SERIES

IF YOU'RE SITTING DOWN right now, stand up and walk a few steps. This simple act—moving from one point to another—was the subject of mathematical and philosophical pondering more than 2,000 years ago by Zeno of Elca. Zeno lived in ancient Greece, purportedly at the time of Socrates, though there aren't many reliable records about his life. Zeno is best known for formulating a series of paradoxes meant to stimulate our thinking about common-sense notions we might have about the world in which we live. The paradoxes touch on motion and time, and therefore incorporate mathematical ideas about infinity.

Zeno's first paradox of motion presents an argument that motion is impossible. Let's assume that you want to walk from your chair to the door. To do so, you must of course reach the midpoint between the two. But before you can reach that point you first must reach another point, the one halfway between the midpoint and your starting position (which is one-fourth of the way to the door). So to traverse a distance of any length you have to cross an infinite number of distances, and since it's impossible to complete an infinite number of tasks, the paradox maintains that you can never reach the door.

This paradox has survived for centuries because it's not clear exactly how to refute it. Because the paradox relies on the notion that space is composed of an infinite number of units, the paradox seems to have been formulated to point out problems with that assumption. Aristotle offered a solution of sorts when he argued that the distance between any two points does not contain an actual infinity of points but instead a *potential* infinity.

Only recently have mathematicians been able to offer another solution. The distance we have to travel to the door can be represented as a convergent series: $\frac{1}{2} + \frac{1}{4} + \frac{1}{8} + \frac{1}{16} + \frac{1}{32} \dots$ Mathematicians have shown that even though this series is infinitely long, it converges to a finite number: 1. In fact, the notion that an infinite series of infinitely small parts can add up to a finite whole forms the basis of Calculus, and allows you to calculate the area under a curve.

Now when you walk to your door, you can appreciate centuries of mathematical reasoning behind your feat!

QUANTUM ZENO EFFECT

Using experiments based on the quantum mechanical properties of atoms, scientists can make atoms freeze in time, in a phenomenon known as the quantum Zeno effect. By observing an atom a certain number of times within a given period, scientists can prevent it from decaying, in essence locking it into a real-life version of Zeno's arrow paradox. (In this paradox, Zeno asks us to consider an arrow shot from a bow. At any particular instant, the arrow occupies a space exactly equal to its length. And since any length of time is made up of a series of instants, Zeno argues, the arrow is always at rest: there is never a time when it is in motion.)

70

Information Theory

MATHEMATICAL CONCEPT: INFORMATION THEORY

EVERY ONCE IN A WHILE, a mathematician comes along who changes the course of history. Claude Shannon was one of those. Shannon, who in the middle of the twentieth century worked at Bell Labs (AT&T's famous research division) and later taught at MIT, was also an electronic engineer interested in communications. His research led to the founding of information theory, which has made possible digital computers, the Internet, and compact discs. He also helped popularize the term "bit," which is short for "binary digit." In other words, Shannon helped make the future possible.

One of Shannon's insights came when he was a master's student at MIT. He realized that the structure of switching circuits in analog computers and telephone networks resembled the structure of Boolean algebra (see entry #77). In a physical way, a closed circuit could represent the logical value "true" and an open circuit could represent "false." In essence, Shannon had realized that you could capture the workings of logic in a machine. You could actually solve problems in logic and mathematics using switches and circuits. This notion led to Shannon's 1938 master's thesis, titled "A

Symbolic Analysis of Relay and Switching Circuits," now called the most important master's thesis of the twentieth century

Later, while working on code breaking during World War II, Shannon became interested in long-distance communication. His thinking eventually became a book, published in 1949, called *The Mathematical Theory of Communication.* Shannon was examining the problems inherent in sending messages long distances over wires: the longer the distance, the more the signal degraded and filled with noise. But by converting the information in the message into basic units called bits, made up of ones and zeros, you could easily reconstruct a degraded message at the other end of a wire, since one and zero are easily distinguished. And the kinds of messages that can be transmitted using those two numbers range from video to photographs to audio files to e-mail: anything that can be transmitted over the Internet.

Shannon also linked bits to the notion of entropy, which for him indicates the amount of information in any particular message. Here is his famous equation:

$$H(X) = -\Sigma p(x) \log p(x)$$

So the next time you send an e-mail, think of Claude.

CIPHERS

In technical language, a cipher is a well-defined method for encoding information. One example is a substitution cipher, in which some letters take the place of other letters in a regular manner. Some substitution ciphers even use multiple alphabets. During the early twentieth century, electromechanical ciphers, such as the German Enigma machine, meant that a machine, not a person, performed the substitutions.

71

Your Social Media Jealousy Has Mathematical Roots

MATHEMATICAL CONCEPT: FRIENDSHIP PARADOX

SOCIAL MEDIA is a big part of society today. Chances are that you use both Twitter and Facebook, and probably Pinterest and Instagram as well. Since the days of Friendster, the popularity of social networks has exploded. But while social media has benefits—it lets you keep in touch with friends and acquaintances from all parts of your life—it also has been found to lower the self-esteem of its users. When people browse their friend networks and see photos of people on exotic vacations, status updates touting promotions or raises, or images of new cars and homes, they can start to feel inadequate: why aren't their lives as good as the lives of their friends?

This phenomenon is a more general case of something known as the friendship paradox. In 1991, sociologist Scott Feld was studying social networks, which in those days did not involve computers and the Internet, and found that in any network of friends, person A's friends would usually have more friends than person A would.

In other words, your friends always have more friends than you do. But how could that be? If I'm your friend, and you're my friend, then we each have one friend. Friendship seems to be balanced that way.

The reason for the paradox is the structure of friend networks. In any network, a few people are more popular than all the rest; they have, on average, a larger number of friends than do the other people in their network. Therefore, the odds are high that a randomly chosen person in that network would be friends with one of those popular people. After all, being popular entails having lots of connections, and you're more likely to be friends with someone who has forty friends than with someone who has two friends. You have a higher probability of being one of the forty than one of the two. But the same principle applies to the majority of the people in the network. The friendship paradox arises because of the very nature of friendships, and a little bit of counting.

What does all this have to do with social media? Well, the friendship paradox does not only apply to face-to-face networks. It also applies to electronic networks. So the odds are high that people you follow on Twitter have more followers than you do, and that most of your friends on Facebook have more friends than you do. And because of recent research by two scientists, the friendship paradox has been extended: not only do your friends have more friends than you do, but they also are probably wealthier and happier than you are. Young-Ho Eom, of the University of Toulouse, and Hang-Hyun Jo, of Finland's Aalto University, analyzed networks of scientists in which each scientist is linked to another if they coauthored a research paper. Eom and Jo found that in any particular academic network, scientist A's connections had more coauthors than scientist A did. They also found that scientist A's connections had more citations and more publications than scientist A did. Eom and Jo worked out the mathematical characteristics of these kinds of networks and learned that if a paradox occurred in a network, it applied to more than one characteristic—not just number of connections or number of citations—if the characteristics met certain criteria. Wealth and happiness fit those criteria.

So the next time you are browsing through social media and feeling bad about yourself, remember that most other people are feeling the exact same way.

SAMPLING BIAS

The friendship paradox is an example of sampling bias. Since your set of friends is biased toward people who have friends in the first place, those friends will most likely have more friends than you do. The sample is automatically more likely to have certain characteristics merely because of the way you've chosen them for study. Another example is the so-called caveman effect. Because people have found many traces of early human beings in caves, it's easy to conclude that our ancestors were primarily cave dwellers. But anything early humans might have left outside of caves would have eroded or washed away. The sample—relics in caves—distorts the conclusion.

72

How an Audio Recording Becomes a Digital Music File

MATHEMATICAL CONCEPT: FOURIER TRANSFORM

WHO KNEW THAT IPODS and math had a powerful connection? It turns out that when you download a song to your computer or play a digital music file on your MP3 player, you are taking advantage of a mathematical equation called the Fourier transform.

The name sounds odd, but think of it as a kind of tool: in essence, it separates complex waves into many simple waves, and it combines simple waves back into one complex wave. And these can be almost any kind of wave, including both sound and light. When sound engineers want to convert an audio recording into an MP3 file, they use the Fourier transform to tease out the individual frequencies of the sound wave and note their amplitudes at each moment. Then, if they want to compress the file to make it easier to transmit over the Internet, they can remove the frequencies that human beings can't hear. On the other hand, the sound that's engraved on a vinyl record is the whole sound wave, with every frequency intact.

The human ear also performs a kind of Fourier transform. At any one moment, a single complicated sound wave is entering the ear, where it vibrates the eardrum and produces electrical waves that the brain then analyzes and interprets. But you never hear that *one* wave: instead, you hear someone on your left talking on their phone, a bus on your right honking at a car, a bird above you chirping and flitting around in a tree. That single wave has been broken apart into its constituent parts and you're now able to identify individual frequencies and sounds and thus better interact with your world.

The Fourier transform also pops up in architecture, especially in earthquake-prone areas. Like every other object, each building in a town or city vibrates at its own natural frequency. Say that a building is located in a city that is struck by an earthquake. If the vibrations caused by the earthquake match the natural frequency of the building, the vibrations will be amplified and the building will have a greater chance of being damaged. (The frequency and strength of the vibrations are two different measurements.) To prevent damage, engineers can use the Fourier transform to analyze the individual frequencies of typical earthquakes at a particular location, and then "tune" a building to make sure that its frequencies don't match those of the earthquakes that tend to happen in that area. Mathematics can literally prevent cities from toppling.

JEAN-BAPTISTE JOSEPH FOURIER

The Fourier transform is named after Jean-Baptiste Joseph Fourier, a French mathematician who lived from 1768 to 1830. He developed it while trying to determine how heat transfers between solid bodies.

73

How Many Colors Do You Need to Make a Map?

MATHEMATICAL CONCEPT: FOUR-COLOR THEOREM

WHETHER YOU'RE A Google Maps aficionado or a devotee of the traditional paper variety, maps are everywhere. They're useful and, despite occasional difficulties with folding, convenient. And they're often very pretty. (Look at maps from the Middle Ages to get a sense of the artistry that used to go into map production.) Maps are also the source of one of the best-known ideas in mathematics: the four-color theorem.

Francis Guthrie, an English math student, first proposed the problem in 1852 while he was trying to color a map of the counties in England. Realizing that he only needed four colors, he wondered whether that same rule applied to *all* maps, even maps that hadn't yet been created. More specifically, Guthrie wondered whether you needed more than four colors to complete a map so that no two touching areas—counties, states, countries, whatever—have the same color. (Two areas touch if they share a border. Sharing a corner, as do Utah and New Mexico, doesn't count.) A proof was finally produced in 1976, 124 years after Guthrie asked his question, by Kenneth Appel and Wolfgang Haken, mathematicians at the University of Illinois at Urbana–Champaign. Though it was a significant achievement, the proof caused a controversy within the mathematics community because it involved the use of a computer.

GRÖTZSCH'S THEOREM

German mathematician Herbert Grötzsch created a proof that is an extension of the four-color theorem: in a planar graph, as long as there are no triangles (essentially no points with three vertices), Grötzsch's theorem states that you only need three colors to achieve the same result.

74

Math Helps Create Your Kids' Favorite Movies

MATHEMATICAL CONCEPTS: GEOMETRY, ALGORITHMS

COMPUTER ANIMATION has advanced a great deal in the past few decades, and no company has been more instrumental to that advance than the animators at Pixar. But computers can only follow instructions that are based on mathematics. So when animators are faced with a new challenge, like portraying the look and movement of Merida's curly hair in *Brave*, they turn to math for help.

Pixar relies on algorithms—sets of instructions—to model complex objects and behavior, and they learned that they would need a whole new set to model Merida's hair, which comprises 100,000 different elements. How tricky is the problem? According to the rules of combinatorics—if there are n elements, there are n^2 ways for them to collide—there are 10 billion ways for all the hair elements to interact.

Pixar has also pioneered a mathematical technique for making sharp edges smooth, crucial for depicting the smooth contours of skin and clothing. Computer animators build up three-dimensional shapes using polygons—shapes of at least three sides—but the resulting object has creases, almost as if it were made out of blocks. By using a process called subdivision, animators find the midpoints of each edge and then average them. Repeated many times, the process transforms a blocky, sharp-edged image into one with life-like curves. Straight lines became parabolas, and Pixar's signature look comes into view.

TOY STORY 2

Pixar's animators and engineers may be smart enough to create algorithms to better animate its characters, but one of its biggest hits, 1999's *Toy Story 2*, was almost lost because of a careless error. The film is one of only a handful to have a perfect 100% score on *Rotten Tomatoes* and also won the Golden Globe for Best Film–Musical or Comedy, but it almost wasn't released because someone accidentally deleted the files from Pixar's computers. This is just a friendly reminder to always back up your files.

75

Candy Crush Saga

MATHEMATICAL CONCEPT: COMPUTER PROGRAMMING

IN THE PAST FEW YEARS, mathematicians have discovered that a popular game being played today on Facebook and mobile devices is actually an example of one of the most difficult problems in the math universe. Math gurus have proved that the game, *Candy Crush Saga*, is a so-called NP problem, meaning that there is no simple, direct way to *solve* it, though the solution is easy to *verify*. NP problems are distinguished from P problems, which can be quickly *solved*.

Computer scientists and mathematicians would love to determine once and for all whether P problems and NP problems are fundamentally the same; that is, whether each problem that one can verify easily is also a problem that one can solve easily. The P versus NP problem has been designated a Millennium Prize Problem by the Clay Mathematics Institute, and whoever determines whether or not P=NP will win a cool million dollars.

Now one of the most popular games on Facebook and mobile devices, *Candy Crush Saga* features a game board with candies of different colors, including yellow lemon drops and red jellybeans.

Players must move candies either horizontally or vertically to create a group of three of the same candy.

PROBLEM REDUCTION

Researchers analyzed the mathematics behind *Candy Crush* in part by using a technique called problem reduction, in which one problem is converted into another one. Problem reduction helps mathematicians determine how difficult a problem is to solve. If the new problem can be reduced to the original problem, then both can be considered equally difficult.

76

Did You Inhale Caesar's Last Breath?

MATHEMATICAL CONCEPT: PROBABILITY

MATH CAN REVEAL basic aspects of human experience that, frankly, boggle the mind. For example, what are the odds that you just inhaled molecules that someone who lived thousands of years ago exhaled in her dying breath? Math can answer that question, and with a surprisingly high degree of precision. How is that possible?

The problem and solution are outlined in *Innumeracy: Mathematical Illiteracy and Its Consequences*, a book by John Allen Paulos, a math professor at Philadelphia's Temple University. Paulos asks if we can determine whether, at this very moment, you inhaled molecules that Julius Caesar exhaled in his last breath as Brutus stabbed him. It turns out, if you accept a few preconditions, the odds are greater than 99%!

1. First you have to assume that the molecules that Caesar exhaled are spread more or less uniformly throughout Earth's atmosphere. (It's been over 2,000 years since his death, after all.)
2. You also have to assume that most of them are still free (not bound to other molecules).

Now, to begin: let's say that there are G (some number) molecules in the atmosphere. And let's say that of those molecules, Caesar exhaled Z (another number) of them. So the probability that you just breathed in one of those molecules is Z/G. Since probabilities are always less than 1, the odds therefore that you did *not* inhale one of those molecules is $1-Z/G$.

Now let's say that you just inhaled three molecules: because of the multiplication principle, the chance that none of these molecules was exhaled by Caesar is $[1-Z/G]^3$. Of course, this principle applies to any number, so in a more general way we can say that if you just inhaled T molecules, the odds that none of them were exhaled by Caesar are $[1-Z/G]^T$

So the probability that you just inhaled at least one of those molecules can be represented by $1-[1-Z/G]^T$. And because Paulos calculated that Z and T are both probably around 2.2×10^{22}, and G is around 10^{44}, the odds come out to be around .99. Amazing.

ASSUMPTIONS

In these Caesar's breath calculations, we made a series of (reasonable) assumptions. Assumptions actually play a large role in mathematics as a whole. For instance, Euclid based his geometrical reasoning on five postulates, one of which is that a straight line may be drawn between any two points. Another is that all right angles are equal.

77

How Do Computers Work?

MATHEMATICAL CONCEPT: BOOLEAN ALGEBRA

COMPUTERS ARE EVERYWHERE. From the smartphone in your pocket to the laptop in your backpack to the giant servers that allow Amazon to process online purchases, computing devices have infiltrated most nooks and corners of everyday life. But how exactly do they work? How do the metal components inside a computer's case let you surf the Internet, share photos with friends, or just add and subtract?

The answer has its roots in math. Computer circuits are built according to principles outlined by George Boole, an English mathematician who lived from 1815 to 1864. Boole became known for applying the methods of algebra to logic, a discipline that focuses on the rules by which someone can draw conclusions based on premises. A classic example of a logical argument—or set of statements that, combined with reason, establish a point—involves Socrates, the ancient Greek philosopher. Here it is:

All men are mortal.

Socrates is a man.

Therefore, Socrates is mortal.

This kind of argument, known as a syllogism, is interesting because as long as the first two statements are true, the third statement *must* be true. And we don't have to use "men," "mortal," and "Socrates." We could replace those terms with anything else. Here's another version:

All birds have wings.

The toucan is a bird.

Therefore, toucans have wings.

But logic can apply to more than simple terms like men and toucans. It can also deal with propositions—statements that can be either true or false. And these statements can be combined using the words "and," "or," and "not." The resulting combinations can have truth-values of their own. Here are some examples of propositions:

There is currently a king of France.

Dogs can breathe underwater.

When a stoplight is red, cars are supposed to stop.

The first two propositions are false; the third is true. Now here are some examples of propositions in combination:

The sun is shining and the cows are grazing on the hill.

Either it's raining or it's snowing.

The car is moving and its wheels are turning.

Let's break down each example:

- In the case of the first combination, if both the sun and cow propositions are true, then the resulting combination is true. If *either* of the smaller propositions is false (or of course, if both are false), then the entire combination is false as well.
- In the second example, the entire combination is true if *either* the rain *or* the snow propositions are true.
- And again in the third example, the combination is true only if both of the smaller propositions are true. If either is false, then the entire thing is false.

Boole's innovation was to notice that one could represent propositional logic arguments using symbols typically used in math. If, for example, the sun proposition above was represented as X and the cows proposition as Y, you could add the propositions together, in a sense, and get a truth value: 1 for true, 0 for false.

The "and," "or," and "not" operations are not just abstract ideas, though. Engineers in the twentieth century learned how to represent them in concrete, physical ways known as logic gates. These gates eventually became incorporated into transistors and computer chips, and underlie all the basic computation that every computer does to this day. All computation is executed based on a particular electrical situation being "true" or "false." So beneath each fancy screen beats the heart of mathematics.

GEORGE BOOLE

Historians maintain that George Boole taught himself Latin as a child. Later in life he became Dean of Science at Queens College, Cork, and married Mary Everest (the niece of George Everest, for whom Mount Everest was named).

78

The Math Behind Birthday Buddies

MATHEMATICAL CONCEPT: PROBABILITY

SOMETIMES MATH REVEALS aspects of the world that seem impossible, but are true nonetheless. Consider, for instance, the birthday paradox. In any group of people, how likely would you say it is that two of those people have the same birthday? The odds, at first, don't seem good since there are 365 days in a year. The chance that a random group would include two people who were born on the exact same day seems profoundly low.

And yet it's not. The probability that two people in a group share a birthday is higher thank you might think. In fact, in a group of only 23 people, the odds are around 50%. How could that be? After all, if you're a member of that group, there are 22 people who might share your birthday, so there are only 22 opportunities for a match. That number doesn't seem like a lot. But remember that you aren't just comparing your birthday to everyone else's. Every other person is comparing birthdays with each other! So, aside from the 22 comparisons to your birth date, there are many, many others.

To see how those arise, imagine all 23 people as dots strung out in a line. (If you like, get a piece of paper and a pencil and draw them out now.) To represent comparisons to person #1's birthday, draw lines from the first dot to each of the other dots. Now do the same for person #2. Notice, though, that the line you draw from person #2 to person #1 is the same as the line you already drew from person #1 to person #2 in the first round of comparisons. Because we don't want duplicate comparisons, the number of comparisons for person #2 is one less than for person #1, which comes out to 21. This process continues: for person #3, the number of comparisons is 20. The *total* number of comparisons is therefore not 22, but 22 + 21 + 20 + 29 . . . , which ultimately equals 253.

Now we come to a principle that often plays a role in mathematical thinking; namely, to establish that something is true, it often helps to prove that its opposite is false. So can we figure out the odds that *no* two people in a group of 23 have the same birthday? Well, remember that a person's birth date is one of 365 possible dates (excluding February 29, which pops up in calendars during leap years). So the chance that any two people have different birthdays is 364/365, since there are 364 possible dates that would be different from the first date. That comes out to a 99.726027% chance that any two people have different birthdays.

Now let's apply this thinking to our group of 23 people. Each comparison has a 99.726027% chance of *not* being a match. Remember that there are 253 possible comparisons in our group, so the overall odds that no two people in the group share a birthday are 99.726027 × 99.726027 × 99.726027 . . . , carried out 253 times. (We can also write this calculation in a shorthand way, as 99.726027^{253}.) The resulting odds are 49.952%. So if those are the odds that no two people in the group have the same birth date, then the odds that some two people in the group *do* share a birth date is 50.048%.

So the next time you're in a large group of people, take a survey of birth dates and see what happens!

SEPTEMBER 16TH

According to Matt Stiles, a data journalist for NPR, September 16 is the most popular birthday for Americans between the ages of 14 and 40. He determined that July and September are the most common birthday months. The least common birthday was February 29, followed by December 25.

79

Bell Choirs and Math

MATHEMATICAL CONCEPT: PERMUTATIONS

THE RINGING OF BELLS brings to mind religious services, college campuses, medieval town squares, and perhaps the perennial Christmas television commercial featuring Hershey's Kisses. But some bell ringing also has a deep connection to mathematics, and particularly to permutations, the rearranging of a given set of objects when the order of each arrangement is important.

The variety of bell ringing that has built-in math is known as *change ringing*, a kind of team activity involving a group of people in which each person is assigned to ring a particular bell. (The number of bells is usually between six and eight, but can range as high as sixteen.) This is the kind of ringing you might have heard in movies after a large wedding, or the crowning of a king. Usually, the bell with the highest pitch is known as the treble, while the lowest-pitched bell is known as the tenor. In any group of bells, the treble is assigned the number 1, and each subsequent bell is assigned a higher number. (If there are four bells, the tenor bell will be number 4.)

In change ringing, all the bells are rung in a particular order called a row, or change, such that no bell is rung more than once per row. And as the rows change, the position of each bell in any row can only change by one place. So the ringers might begin by ringing the bells in this order: 1, 2, 3, 4. As they continue, they might ring 2, 1, 4, 3 and then 2, 4, 1, 3. In addition, each order must not be repeated. At the end of the ringing session, the ringers return to 1, 2, 3, 4. If you live in North America, or would like to hear change ringing for yourself, go to the website of the North American Guild of Change Ringers at *www.nagcr.org*.

CARILLONS

Change ringing differs from other kinds of bell ringing such as playing the carillon, in which a musician sits at a chair or bench and presses a series of levers set up like a keyboard. With 77 bells, the largest carillon in the world resides at Kirk in the Hills Presbyterian Church, in Michigan.

80

Bayesian Statistics

MATHEMATICAL CONCEPT: BAYESIAN STATISTICS

IF YOU ASKED a college student to name the area of mathematics that was the most dull, the most boring, the most devoid of any redeeming qualities, she might choose statistics. The word alone conjures up images of calculators and tables upon tables of dry numbers. At least, this is what the stereotype calls to mind.

What if I told you that statistics were nowhere near as mind-numbing as you might think?

One way to convince you would be to talk about Bayesian statistics, a discipline introduced by Thomas Bayes, a Presbyterian minister who lived in England in the 1700s. The kind of statistics you might be more familiar with is known as frequentist statistics. If you were playing blackjack and had been dealt a nine and a king, you could use frequentist statistics to figure out your odds of getting blackjack on the next deal of the cards.

Bayesian statistics, on the other hand, constantly revises the odds as new information comes in. In the case of playing a game of blackjack, you wouldn't just consider the odds of getting a three, or analyze bare data. You would also take into account which cards had already been dealt and the skill of the dealer. With each new nugget of information, the odds of the outcome are revised.

However, Bayesian statistics can do more than calculate the odds of getting a winning hand in a card game. It can save lives. For example, it was used to help locate John Aldridge, a fisherman who fell from his lobster boat off the coast of Long Island in 2013. He was lost in a vast area of the Atlantic Ocean, but as the Coast Guard took into account the currents in the area, as well as the routes that the rescue helicopter had already taken, they were able to narrow down the fisherman's possible whereabouts. The Coast Guard estimated the time at which Aldridge most likely fell off the boat, and the computer program they used—SAROPS—analyzed wind and ocean currents to find his most likely locations in the ocean. When they eventually found him, he had been afloat for 12 hours.

BAYESIAN INFERENCE

The classic example of a Bayesian inference is a newborn baby watching the sun rise. With each morning that the baby observes the rising sun, the baby gets more and more evidence that the sun, in fact, rises in the morning, and therefore has better and better reason to believe that the sun will rise in *future* mornings. Updated information, in the form of new observations, gets factored in to whether the baby should expect future sunrises.

81

Baseball and ERA

MATHEMATICAL CONCEPT: STATISTICS

PERHAPS NO SPORT is as math-heavy as baseball. Statistics permeates every aspect of the game, from hitting to fielding to pitching, and no serious baseball scholar can go without at least a rudimentary knowledge of numbers. But how do people calculate these numbers?

Let's examine ERA, or the earned run average. This statistic only applies to pitchers, and is meant to capture how good they are. In baseball's early days, there were no relief pitchers; the pitcher who started the game was expected to finish it. So if someone wanted to figure out how effective a pitcher was at striking out batters, or at least making sure that they didn't make it on base, they could just look at how many games the pitcher won. But when relief pitchers began appearing, the outcome of a game began to depend on more than one pitcher, so just totaling wins and losses would not accurately capture a particular pitcher's skill.

The earned run average solves the problem: it focuses on innings, not whole games. To calculate an ERA, all you have to do is add up all the runs that were scored on a pitcher and divide that number by the number of innings that the pitcher has pitched.

(Earned runs are those that are the fault of the pitcher, not caused by errors made by other players.) Then you multiply that number by 9, the number of innings in a game. For example, if a pitcher allowed 30 runs over 90 innings, his ERA would be 3.00.

What is considered a good ERA has varied throughout the course of baseball's history, but the lower the better. In the early 1900s, a good pitcher would have an ERA under 2. These days, an ERA under 4 is considered respectable.

CLAYTON KERSHAW

From the 2011 through the 2014 Major League Baseball seasons, Clayton Kershaw of the Los Angeles Dodgers had the lowest ERA of all active pitchers, with his best in 2014 at 1.77. By way of comparison, Tim Keefe of the Troy Trojans has the lowest ERA in baseball history, at 0.86 in the 1880 season.

82

Bacterial Division

MATHEMATICAL CONCEPTS: KNOT THEORY, SHAPES, DIVISION

IN EVERY PERSON'S LIFE, nothing is more certain than death and taxes, except perhaps the presence of bacteria. These tiny organisms live everywhere, on every continent and in every environment, even in our guts, where they help us digest our food. Because they are minuscule, a crucial tool in the study of bacteria is the microscope. But another tool is mathematics, which can clarify aspects of the bacterial life cycle and therefore contribute to human health.

Unlike human DNA, bacterial genetic material is not organized in a double helix shape. Instead, it's shaped like a circle. When a bacterial cell replicates, dividing itself into two daughter cells, its DNA also must divide in two. And as it divides, it tangles and untangles, knots and unknots, until two new circles of DNA are produced. Biologists have had a hard time figuring out the exact workings of the replication mechanism, but mathematicians may have come to the rescue.

Using a mathematical tool called tangle analysis, researchers were able to get a better sense of how enzymes—molecules that start and stop chemical reactions—both link and unlink the circles of DNA, including the shapes that the molecules take. (Experiments had given scientists a sense of the process, but had not made clear the specific steps.)

The more that scientists understand how bacteria reproduce, the better prepared they'll be to create new generations of antibiotics. Mathematics could very well be the reason you recover more quickly from your next illness!

MICROBES

Scientists continually study the microbes in our bodies and how they keep us alive. The human body has about 100 trillion cells, but only 10% of those are estimated to make up your body. The rest are bacteria, viruses, and other microorganisms.

83

Astrolabes

MATHEMATICAL CONCEPT: STEREOGRAPHIC PROJECTIONS

STEREOGRAPHIC PROJECTIONS aren't just found on wall maps (see entry #28). For hundreds of years, they were the basis of one of the most popular astronomical instruments in human history: the astrolabe. Often made of brass and measuring at least 6 inches in diameter, the astrolabe was a kind of portable computer that would let sailors make crucial calculations about the time of day or night, the altitude of celestial objects over the horizon, the times of future sunrises and sunsets, and the latitude of the observer. They could also be used to make astrological calculations. (Astronomy was linked to astrology for hundreds of years.)

Astrolabes are one of the oldest scientific instruments. They were known in ancient Greece, and astrolabe technology was preserved by the Islamic world during the Middle Ages. In fact, they continued to be used during the Renaissance and up until the 1700s, when the sextant become more popular. (The mirrors in sextants allowed navigators to make calculations based on Earth's actual horizon, not the false horizon that astrolabe users had to rely on.)

Astrolabes could make complex calculations, but they only consisted of a few parts:

- The *mater* was the circular metal frame that held all the other parts.
- The *climates* were metal plates with etched lines that would fit into the mater. Different climates could be attached to the mater, based on where in the world the astrolabe-user was located.
- The *rete* was secured to the top of the climate, and included cutouts so the observer could see through to the climate below. The rete also included markers that pointed out important features on the climate.
- Astrolabes also included an *alidade*, which helped the observer figure out the altitude of celestial objects.
- Astrolabes came with a ring affixed to the top edge, so the device could be suspended by string (aiding the calculations).

The stereographic projections can be seen on the climates. Each climate had an etched pattern of lines that corresponded to lines of latitude on Earth, as if a collapsible globe had been flattened on top of the plate. Some climates included lines representing other map features, like hour lines, lines of azimuth, and almucantars. In short, stereographic projection made possible the navigation of the oceans, as well as the navigation of one's own astrological destiny. And all because of math.

ASTROLABE WATCHES

Show your math geekery by wearing an astrolabe watch (available online), though it will probably be too small to be useful or practical!

84

Angle of Repose

MATHEMATICAL CONCEPT: ANGLE OF REPOSE

YOU CAN FIND MATH almost everywhere, including your dinner table. Pour some table salt into a pile on a sheet of paper and it will form a cone. But this cone isn't just pretty. It also demonstrates a phenomenon known as the angle of repose. This is the angle that the surface of the pile makes with respect to the horizontal surface of the table.

In fact, all granular materials—including sand and rocks—have angles of repose, even boulders that tumble down mountainsides during avalanches. Moreover, the angle isn't random, and it doesn't change with each occurrence. It depends on a combination of factors, including the size of the particles, whether they're smooth or jagged, whether there's water between the particles (which might make them stick together), and how rough the underlying surface is.

The angle of repose of table salt is 32 degrees, but angles of repose can go higher: 45 degrees for tree bark and coconut flakes. An angle of repose can also go lower: wet clay has an angle of repose of 15 degrees. People can even use angles of repose to calculate whether a pile of material, like gravel, is likely to collapse.

So feel free to dump out your salt shaker at your next family gathering. Tell everyone you're doing it for math!

ANGLES OF REPOSE OF VARIOUS SUBSTANCES

ITEM	ANGLE OF REPOSE
Ashes	40 degrees
Bran	30–45 degrees
Gravel	30–35 degrees
Dry Sand	34 degrees
Snow	38 degrees
Wheat	27 degrees

PART 4
SPECIAL NUMBERS

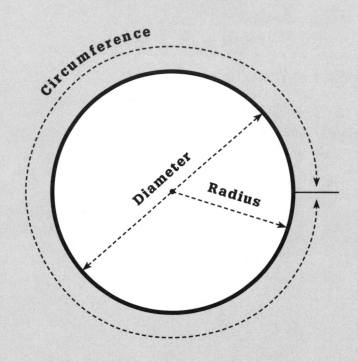

$$\frac{\text{Circumference}}{\text{Diameter}} = \pi = 3.14159$$

85

What's All the Fuss about Pi?

MATHEMATICAL CONCEPT: IRRATIONAL NUMBERS

TAKE ANY CIRCLE and measure both its circumference (the distance around its edge) and its diameter (the distance from one side of the circle to the other, measured as a line that passes through the circle's center). Are they the same size? If not, how much larger is one than the other?

It so happens that the circumference is always larger than the diameter. That alone is an amazing fact. For any circle in the world— the rim of your coffee cup, a wheel on your bike, a nickel—the line tracing the outside is always larger than the line going through the center. You don't have to measure to make sure (though you could, just to prove to yourself that what I'm claiming is true). This property is universal; it applies to all circles, everywhere, for all time. (I'm assuming here that the circles we are discussing are all on flat surfaces.)

Now we come to the second amazing part of the relationship between a circle's circumference and diameter. For all circles, the circumference is always larger by the *same amount*. That amount isn't a fixed number, like 39: the absolute number difference between the circumference and diameter of a large circle will, of course, be larger than that of a small circle. What stays the same is the *ratio*, or the comparative difference. "That's great," you say. "How much larger *is* the circumference? Is it twice as large? 1.5 times as large?"

Here's where circles get weird. In a sense, it's *hard to say* exactly how much larger the circumference is. People have known for thousands of years that the circumference is about three times larger than the diameter, but in fact the number is more like 3.14. An even more precise number would be 3.14159. However, that series of numbers after the decimal point goes on *forever* without repeating. To date, the most precise calculation of this ratio extends to eight quadrillion places after the decimal.

This number—the ratio of a circle's circumference to its diameter, or, put another way, how much larger one is than the other—is known as *pi*, after a letter of the Greek alphabet. The name doesn't matter, though: we could just as well call it "Frank" or "Sam" or "Felicia." What matters is how pervasive it is in our world—it's built into every circle—and how strange it is. To grasp further how strange pi is, imagine that your friend asked you how much taller you were than your dog. What if you answered, "Well, I'm not quite sure. I'm about twice as tall as my dog, but the more I measure the more I realize I'll never get the exact number." How could there not be a definite answer to the question? That's the baffling nature of pi.

PI DAY

For math geeks, March 14 is a special date. It's known as Pi Day, and celebrations begin at 1:59 P.M. (Combining the numbers associated with the month, day, and time yields 3.14159.) Pi Day began at the Exploratorium, San Francisco's hands-on museum of science, art, and human perception.

Primes

MATHEMATICAL CONCEPT: NUMBER THEORY, PRIME NUMBERS

SOME NUMBERS ARE SPECIAL, and some of the most special numbers are the primes. Prime numbers are divisible only by themselves and 1. For instance, 5 is a prime number, since it's only divisible by 5 and 1. But 10 is not prime, since it's divisible by 1, 2, 5, and 10. Prime numbers have fascinated mathematicians for more than 2,000 years, since the time of the ancient Greeks and Euclid (one of the greatest mathematicians in history, and author of *Elements,* one of the most influential books in human civilization).

Primes are interesting because they seem to be the fundamental constituents of all the other numbers—in fact, prime numbers are sometimes called the atoms of mathematics—but the way in which they appear doesn't seem to obey any law. According to the fundamental theorem of arithmetic, every number greater than 1 is either a prime or can be produced by multiplying together a series of primes. Here are some examples:

2 is prime.
3 is prime.

$4 = 2 \times 2$

5 is prime.

$6 = 2 \times 3$

7 is prime.

$8 = 2 \times 2 \times 2$

$9 = 3 \times 3$

$10 = 2 \times 5$

11 is prime.

$12 = 2 \times 2 \times 3$

13 is prime.

And so on.

How many primes are there? Euclid proved that there are, in fact, an infinite number of them. No matter how far out we travel on the number line, we will never get to the last prime. There will always be more.

The way Euclid came to this conclusion is worth examining, since it's a good example of how mathematicians use reasoning to learn about numbers and their properties.

1. First, remember that every number is either a prime or can be rewritten as a product of only primes.

2. Second, we're going to use a particular kind of proof technique called reductio ad absurdum: we're going to assume the opposite of what we're trying to prove. If we find that this opposite claim can't possibly be true, then we know that *its* opposite—what we were originally wanting to prove—must itself be true.

In other words, we're going to proceed by assuming that there are a limited number of primes. Spoiler alert: if we find that our proof hits a logical snag, we will have indirectly shown that the opposite case is true and that the number of primes is unlimited.

Step one: let's invent a number. Let's call the number "George." Let's say that we can get George by multiplying all the primes together, first through last, and then adding 1. (Remember that we are assuming that there are a limited number of primes.) We know that George must be either prime or the product of two primes. We can see right away that if George is prime, we have established that there is a prime—George!—not on our original list. We can now stop and pat ourselves on the back, since our result would hold for any size list of primes.

But let's consider the other case: let's say that George is *not* prime. That means that George must be the product of two or more primes. But none of the primes on our list would suffice, since if we use them in our calculations we'll always end up with a remainder of 1. Therefore, there must be some *other* primes not on our list, which, when multiplied together, result in George. Again, we have established that for any given set of prime numbers, there will always be prime numbers outside it.

This is just one example of the power and beauty of mathematical reasoning.

FERMAT PRIMES

Some primes are more exotic than others. A Fermat prime, for instance, is a Fermat number, which has the form $2^{2n} + 1$, that is prime. The only known Fermat primes, though, are those for which n equals 0, 1, 2, 3, and 4, equaling 3, 5, 17, 257, and 65,537, respectively.

87

Internet Security

FORGET E-MAIL AND SOCIAL MEDIA. The Internet's most profound impact on the world might very well be online shopping. But eCommerce seems fraught with peril. When you type your credit card number into a site like Amazon and click "Purchase," what stops cyberthieves from intercepting that transmission and stealing the number?

Enter mathematics. Internet security, and public-key cryptography in general, is based on prime numbers, a special kind of number that is divisible only by itself and 1. (By comparison, the number 10 is divisible by itself, 1, 2, and 5.) Examples of primes include 1, 3, 5, 7, and 11, but there are an infinite number of primes, and accordingly they can grow to huge lengths. There's a mathematical proof that shows *why* there are an infinite number of primes, but that's another story (see entry #86).

When you submit your credit card number online, you take advantage of something called the RSA algorithm. Named after its inventors—Ron Rivest, Adi Shamir, and Leonard Adleman—the algorithm, created in 1977, is an encryption system that relies on prime numbers of enormous length. Basically, the system secures your private data by transforming your credit card number into another, extremely large number, one that is produced by multiplying two, large, randomly chosen primes. You could transform that new number back into the credit card number if you knew the two primes, but the security of your data is assured by another mathematical fact: it is profoundly difficult to factor a large number into two primes. In fact, for a number of sufficient length, it could take a network of supercomputers hundreds or thousands of years to find the two correct primes.

Over the past few years, though, there have been rumblings that the RSA system is not as ironclad as once believed. The system's integrity depends on the generation of random prime numbers, but the programs that do so—known as random number generators—seemingly do not always create perfectly random numbers. That discrepancy leaves open the possibility that crooks could discover the identity of the two primes and steal sensitive information.

For now, eCommerce—as well as online banking and communication—seems more or less safe, but let's hope that mathematicians somewhere can devise a new, more secure system.

BITCOINS

Anyone who uses Bitcoin—the virtual currency invented in 2008—has to have a kind of digital wallet, which remains secure because of public-key cryptography. In fact, Bitcoin is the first currency to rely on cryptography for its implementation. In order to access your Bitcoins, you must have two keys; one is public and the other is secret, like a pin number or a password. The two keys are mathematically linked.

88

The Wonder and Frustration of Infinity

MATHEMATICAL CONCEPT: INFINITY

EVERYONE HAS SEEN the symbol for infinity: a figure-eight symbol lying on its side (∞). But what is infinity and what is its relationship to mathematics?

Infinity is sometimes understood as an impossibly large number, but that notion isn't quite accurate. Infinity isn't really a number; instead, it's an idea. It's the concept of the unlimited, never-ending, and unbounded. Nevertheless, it appears in math again and again. We say that the sequence of numbers in pi goes on to infinity; so do the numbers you get when you divide 1 by 3. In geometry, we say that there are an infinite number of points on a line, and that lines extend in both directions forever without end. Infinity is both a native and an alien of the mathematical world.

Infinity also appears in art. M.C. Escher depicted ants crawling along a Möbius strip in a seemingly never-ending journey, while in a short story titled "The Library of Babel," author Jorge Luis Borges imagined a repository of books without end, containing every possible permutation of letters and punctuation, a collection that would necessarily contain every book that has been written, and that will be written in the future.

The notion of infinity has also led to some odd-sounding ideas. In the late nineteenth and early twentieth centuries, mathematician Georg Cantor reasoned that there could be infinities of different sizes. Both the natural numbers (1, 2, 3, 4 . . .) and the real numbers (which include numbers like pi, ⅓, and 45.6778765) are infinite, but the infinity of the real numbers is *greater* than the infinity of the naturals. Thinking about infinity leads to other counterintuitive notions. You might think that the number of points in a line that's 1 foot long is less than the number of points in a line of infinite length, but in fact both lines contain an equally infinite number of points.

Infinity is fascinating, but, like much related to mathematics, also twists the mind in baffling and frustrating ways.

FINITISM

Not all mathematicians accept the notion of infinity. One branch of the philosophy of mathematics is known as finitism; its adherents maintain that only finite objects are real. As mathematician Leopold Kronecker put it, "God created the natural numbers. All else is the work of man."

89

Fibonacci Numbers in Nature

MATHEMATICAL CONCEPT: FIBONACCI SEQUENCE

IN 1202, an Italian mathematician published *Liber Abaci*, a book that included what has come to be recognized as a magical sequence of numbers. Known as the Fibonacci sequence (after the mathematician, whose name was Leonardo Fibonacci), the numbers seem innocuous at the beginning. The sequence begins with 1 and 1 (or sometimes 0 and 1), and each successive number is generated by adding the previous two numbers. So, since 1 + 1 equals 2, 2 is the next number, followed by 3, 5, 8, 13, 21, 34, and so on. One interesting feature of this sequence is that the numbers get large relatively quickly. (Calculating the sequence to the eighteenth position yields the following numbers: 0, 1, 1, 3, 5, 8, 13, 21, 34, 55, 89, 144, 233, 377, 610, 987, 1,597, 2,584.) But another aspect of the Fibonacci sequence is far more noteworthy: to an astonishing degree, the Fibonacci numbers seem to be embedded in the natural world.

It seems that if you examine the world with mathematics in mind, you'll find Fibonacci numbers everywhere. You can find them, for instance, in the way that leaves grow. The arrangement of leaves on a stem is called phyllotaxis, from Greek words meaning plant and arrangement. On some plants, leaves sprout from a stem in a spiral pattern. If you begin with one leaf and go up and across the stem along that spiral, you can count how many leaves it takes to get to a position on the stem directly above that first leaf. The relationship between the number of leaves you meet as you go around the stem from one leaf to the next right above it and the number of times you go around the stem is made up of two Fibonacci numbers, and is known as the phyllotactic ratio. For example, the ratio of apple leaves is 2:5, while for blackberry and hazel plants the ratio is 1:3.

Or examine a pinecone. If you look down at the pinecone's top, you'll be able to distinguish two sets of curves in the cone's scales: one circling in a clockwise direction, the other going counterclockwise. If you count the number of both sets of curves, you'll find that they are adjacent Fibonacci numbers.

HONEYBEES AND FIBONACCI

Fibonacci numbers also appear in the family trees of honeybees. Since male honeybees develop from unfertilized eggs, they only have one parent (the queen). Female honeybees, on the other hand, develop from eggs that have been fertilized and thus have two parents: the queen and a male. If you analyze the ancestors of both male and female honeybees, the Fibonacci numbers become apparent: male bees have 1 parent, 2 grandparents, 3 great-grandparents, 5 great-great-grandparents, 8 great-great-great grandparents, and so on. Female bees have 2 parents, 3 grandparents, 5 great-grandparents, 8 great-great-grandparents, etc.

Dewey Decimal System

MATHEMATICAL CONCEPT: GENERAL NUMBERS

NUMBERS ARE ALL AROUND US, and they perform a wide variety of roles. Some, like the numbers on traffic signs, tell people how quickly they are allowed to drive. Others, like the numbers in scoring systems in sports, help us keep track of which team is winning. Those used in medical measurements help us quantify aspects of our health (like blood pressure and cholesterol level) and to think more concretely about parts of our world that are hard for us to sense. And some numbers are used to classify and organize. You can see examples of these numbers on the spines of library books, and they belong to the Dewey decimal system.

Formulated by Melvil Dewey and published in 1876 while he was working in the library at Amherst College, the system revolutionized library science. Before the introduction of Dewey's system, library books were often placed on shelves according to the date when the library had acquired them. Instead, Dewey made it possible for librarians to shelve books according to subject, making it easier for library patrons to browse the shelves and find items on their own. (Previously, library shelves had been closed to patrons;

only librarians were allowed to retrieve books.) Following are the ten broad categories that comprise the Dewey decimal system:

000–099: general topics
100–199: philosophy and psychology
200–299: religion
300–399: social sciences
400–499: language
500–599: natural sciences and mathematics
600–699: technology
700–799: arts
800–899: literature and rhetoric
900–999: history, geography, and biography

510

It's likely that the book you're reading right now will be classified in Dewey as 510, for the Mathematics section of the library.

91

Random Numbers: Are They Really Random?

MATHEMATICAL CONCEPTS: NUMBER THEORY, CRYPTOGRAPHY

EVERYONE HAS HEARD someone say, "That's so random." But what *is* randomness, anyway? And what does mathematical randomness have to do with our everyday lives?

It turns out that in the age of the Internet, randomness is crucially important. Online transactions—involving banks, retailers, and other organizations that require the transmission of sensitive information—rely on generating random numbers to create a secure connection (see entry #87). A series of numbers is random when there is no discernible pattern, no way to predict which number will appear after any other number in the sequence. A roll of the dice is random, but because the volume of online transactions is so large and the demand for random numbers so intense, rolling dice—or pulling numbers out of a hat or using a shuffled deck of cards—just won't do.

Scientists have turned to computers to generate random numbers, but because computers, at heart, are deterministic machines (they follow rules), computer-generated random numbers are not truly random. If someone were to learn the algorithm by which the computer picks numbers, as well as the seed, or beginning value, that person could in theory predict numbers in the sequence. The sequence appears to be random, but in truth is not. That's why computer-based random number generators are also known as pseudorandom number generators.

However, scientists have moved closer to producing truly random numbers using devices that examine physical phenomena like radio noise, the quantum behavior of photons, and the emission of heat. Those phenomena can determine their own randomness without a person-created algorithm. With our reliance on randomness increasing along with our reliance on the Internet, we need all the truly random numbers we can get.

RANDOM NUMBERS AND THE LOTTERY

Random number generators aren't just tools for scientists and mathematicians. If you like playing the lottery, you can take advantage of several websites that generate random lottery picks. Just be aware that that won't necessarily get you any closer to a winning number.

92

Powers of Ten

MATHEMATICAL CONCEPT: SCALE

IN 1977, designers Charles and Ray Eames produced a film demonstrating the vast range of scales in our world, from the realm of the tiny atom to the boundaries of the known universe. Called *Powers of Ten: A Film Dealing with the Relative Size of Things in the Universe and the Effect of Adding Another Zero*, the movie begins with a couple relaxing in a Chicago park. The camera is set at a position about one meter above the couple, and soon begins to zoom out: every ten seconds the camera moves ten times farther away from the couple, and the frame shows ten times as much. Soon, the viewer's perspective grows to show all of Chicago, then all of Earth. Eventually the viewer leaves the solar system and even the Milky Way galaxy, traveling in jumps of light years to view the structure of galaxies populating the universe.

The perspective then returns to the couple and starts zooming in on the hand of the man. This time the camera moves ten times inward every ten seconds, eventually making its way to the protons and neutrons in the nucleus of a carbon atom in the man's hand.

In all, the journey encompasses forty powers of ten, from 10^{15} meters to a distance of 10^{24} meters, a journey made possible simply by adding a zero to the end of our scale measurement.

GOOGOL

Some numbers can be named but are far too large to comprehend. One of those, a googol, is 10^{100}, or 1 with 100 zeroes after it. The term "googol" was created in 1938 by Milton Sirotta, the nine-year-old nephew of mathematician Edward Kasner. How big is a googol? Consider that the total number of elementary particles in the universe is thought to be around 10^{80}.

93

Metric System

MATHEMATICAL CONCEPT: MEASUREMENT

FOR MOST OF THE WORLD, the go-to measurement standard is the metric system. Invented in France in 1799, the metric system was designed to replace the hodgepodge of different measuring standards in Europe that varied from town to town, county to county, and country to country. The metric system simplifies measurement by defining a set number of base units—like the meter, the kilogram, and the second—and allows people to modify them by adding a prefix. And because the prefixes are based on multiples of 10, the metric system is known as a decimal system. Some of the prefixes are as follows:

micro = 1/1,000,000
milli = 1/1,000
centi = 1/100
deci = 1/10
deca = 10
hepta = 100
kilo = 1,000
mega = 1,000,000

So, for example, a kilometer is a distance equal to 1,000 meters, and a milligram is an amount of mass equal to 1/1,000th of a gram. Other metric base units include the ampere, for electric current; the kelvin, for thermodynamic temperature; the mole, for amount of substance; and the candela, for luminosity.

One of the fascinating aspects of the metric system is how, exactly, the base units are defined. Let's consider the meter. At first, the standard meter was defined as one 10-millionth of the distance between the Equator and the North Pole, as measured through the city of Paris. But left out of the calculations was the slight flattening of Earth's curvature, meaning that the calculation was off by .2 millimeters. (Earth is not a perfect sphere.) Later, in 1889, the meter was defined as the length of a particular metal bar made of platinum and iridium. Scientists were uncomfortable with this standard, though, because the bar could change shape in different temperatures and deform if not supported correctly; any attempt to define the meter bar's temperature and support system would have to rely on metric units, including centimeters. But since the meter bar was supposed to be the standard metric unit of length, defining it in terms of other metric units of length wouldn't do.

These days, a standard meter is defined in a much more universal way: as the distance light travels in a vacuum in 1/299,792,458 of a second.

THE ENGLISH SYSTEM

Opposed to the metric system is the English system, with units of measurement that developed in England over the centuries and incorporated measurements that were used by the ancient Romans and Anglo-Saxons. The names of these units tend to have a nice sound, and include terms like "dram," "gill," "hogshead," and "rundlet." And of course, if you're familiar with wine, you know that a 4.5-liter bottle is known as a Rehoboam, while a 15-liter bottle is known as a Nebuchadnezzar.

94

Attoseconds

**MATHEMATICAL CONCEPTS: SMALL NUMBERS,
MEASUREMENT**

AT ITS MOST BASIC LEVEL, mathematics deals with numbers, and some of those numbers are quite simply inconceivable. For instance, what is the shortest duration of time that human beings have ever measured?

Recently, scientists at Germany's Max Born Institute for Nonlinear Optics and Short Pulse Spectroscopy found a way to measure events in increments of 12 attoseconds. How short is an attosecond? One of them is a millionth millionth millionth—or 1/1,000,000,000,000,000,000—of a second. How short is that? In one attosecond, light can travel the length of three hydrogen atoms. Another way to think about this astonishingly short amount of time is to make an analogy: an attosecond is to one second as one second is to 32 billion years (almost three times the age of the universe).

The prefix *atto-* means eighteen in Danish, and has become part of the metric system's order of magnitude. The metric system has a large range of prefixes that indicate quantities on both small and large scales. Here is the current prefix list:

yotta-	10^{24}, or 1 septillion
zetta-	10^{21}, or 1 sextillion
exa-	10^{18}, or 1 quintillion
peta-	10^{15}, or 1 quadrillion
tera-	10^{12}, or 1 trillion
giga-	10^{9}, or 1 billion
mega-	10^{6}, or 1 million
kilo-	10^{3}, or 1,000
milli-	10^{-3}, or 1,000th
micro-	10^{-6}, or 1 millionth
nano-	10^{-9}, or 1 billionth
pico-	10^{-12}, or 1 trillionth
femto-	10^{-15}, or 1 quadrillionth
atto-	10^{-18}, or 1 quintillionth
zepto-	10^{-21}, or 1 sextillionth
yocto-	10^{-24}, or 1 septillionth

THE FLASH

The Flash, a comic book character who can run at the speed of light, can perceive events that last for less than one attosecond, which of course appear in less time than a blink of an eye for us.

95

Golden Ratio in Art and Architecture

MATHEMATICAL CONCEPT: GOLDEN RATIO

WHAT DO LEONARDO'S *The Last Supper* and Michelangelo's Sistine Chapel have in common? They both incorporate something called the Golden Ratio, a relationship between numbers that is found throughout both nature and human endeavor.

If you recall from math class, a ratio is a way to compare two numbers or measurements. For example, let's say you have a four-door sedan. It will have four seats and, most likely, four wheels. The ratio of doors to wheels, therefore, is 4:4. (We could then reduce the ratio to 1:1.) Or let's say you love pets and have 2 dogs and 5 cats. The ratio of dogs to cats in your home would then be 2:5.

The Golden Ratio is just like these kinds of ratios, except that instead of 1:1 or 2:5, the Golden Ratio is 1:1.618. (The second number actually goes on to infinity without repeating; I just shortened it for convenience. Extended to more decimal places, it looks like this: 1.61803398874989. . . .)

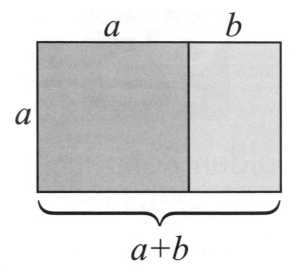

$a+b$

You can also find the Golden Ratio in a shape called the Golden Rectangle. The ratio of the width to the length of this rectangle is 1:1.618 (extended to infinity). In addition, if you create a square inside this rectangle, the leftover rectangle has the same proportions as the original, larger rectangle! In other words, the ratio of the smaller width to the smaller length (b:a) is the same as the ratio of the larger width to the larger length (a:a+b).

Architects and artists have used the Golden Ratio in their designs for centuries, since human beings seem to find that proportion especially pleasing. Here are some examples:

- An imaginary rectangle drawn around the front of the Parthenon in Athens has just those proportions.

- The Great Pyramid of Giza, one of the original Seven Wonders of the World, also incorporates the Golden Ratio. The ratio of the length of one of the sides of the Great Pyramid—a slant—to one-half of the length of the Pyramid's base equals 1.61804.

- If you analyze Leonardo da Vinci's *The Last Supper*, as well as Michelangelo's *The Creation of Adam*, on the ceiling of the Sistine Chapel, you'll notice that the compositions of both rely on the Golden Ratio.

Any museum will be full of examples of 1:1.618, as will many cityscapes. The numbers are all around you.

GOLDEN RATIO: FACT OR FICTION?

Many people believe that people have used the Golden Ratio in art and architecture for millennia. Some mathematicians, on the other hand, argue that there is no evidence for this claim, and that any notion that the Golden Ratio was used to create the Great Pyramid, the Parthenon, or even Leonardo da Vinci's work is a myth.

96

The Golden Ratio in Your DNA

MATHEMATICAL CONCEPTS: GOLDEN RATIO, FIBONACCI SEQUENCE

BESIDES OCCURRING throughout the world of art, the Golden Ratio is a recurring theme in nature. In fact, the Golden Ratio is part of *you*. It's built into every DNA molecule in every cell in your body. (DNA, or deoxyribonucleic acid, encodes instructions for building proteins, which in turn help make up the structure of the body's organs and tissues and regulate their functions. Hormones and enzymes are proteins, as is actin, which helps muscles contract. Actin also is part of a cell's interior skeleton, which gives a cell its shape.) The structure of the DNA molecule was famously revealed by James Watson and Francis Crick in 1953 to be a double helix. Each complete twist of the helix is 34 angstroms long and 21 angstroms wide—an angstrom is 100 millionth of a centimeter—and the ratio of those two numbers is 1:1.6190, very close to the Golden Ratio's number of 1.618.

Those numbers, 34 and 21, are special for another reason: they occur in the Fibonacci sequence, a series of numbers described by

Leonardo Fibonacci in the thirteenth century (see entry #89). Each number in the sequence is determined by adding the two previous numbers. And when you compare any two neighboring numbers in the sequence, the ratio between the two approximates the Golden Ratio. Furthermore, the higher in the sequence you go, the closer the approximation gets. So, the ratio between 5 and 8, two early numbers in the sequence, is 1:1.6, while the ratio between 377 and 610, two later numbers, is 1:1.61803714. That result matches the Golden Ratio out to five decimal places.

PHI AND THE GOLDEN RATIO

The Greek letter phi Φ was first used to symbolize the Golden Ratio in 1909 by American mathematician Mark Barr in honor of Phidias, an ancient Greek sculptor who many people believe used the ratio in his work.

97

Using a Kid's Toy to Make Epitrochoid Curves

MATHEMATICAL CONCEPT: SHAPES

SOME OF THE MOST INTERESTING mathematical shapes show up in toys. For instance, you can get a good sense of epitrochoids, as well as hypotrochoids and roulettes, by messing around with a spirograph. If you've ever had a chance to play with this drawing toy, you know that it consists of two plastic circles that have open middles, like donuts, with gear teeth on both the inner and outer edges. It also comes with a series of solid plastic wheels with toothed outer edges and holes punched somewhere in the wheels' interiors that accept the tip of a pen. You secure one of the two plastic circles to a sheet of paper with a cardboard backing, and then you choose a wheel, insert a pen tip in the hole of the wheel, and place the wheel against either the circle's inner or outer gear teeth. By using the pen to move the wheel around the circle, you can draw intricate geometrical patterns.

Here's where the odd math terms come in. If you place the wheel on the *outside* of the circle, the pattern you trace is an *epitrochoid*. It looks like a series of curves that swoop toward and then away from the circle's edge. On the other hand, if you place the wheel on the *inside* of the circle, the pattern you trace is a *hypotrochoid*, which can look like a starfish, a pentagon with bowed-in sides, or a star. Both curves are special examples of a roulette, a kind of curve formed when a shape rolls along a fixed surface, and a point on that shape traces out a wobbly line. (The shape doesn't have to be a circle.)

WANKEL ROTARY ENGINES

The housing that holds Wankel rotary engines, which power some Mazda cars, is shaped like an epitrochoid. The engine was invented by Felix Wankel, a German engineer who received his first patent for the device in 1929. Unlike piston-driven engines, the Wankel engine has just one moving part: a rotating piece shaped like a triangle with slightly curved edges.

98

The Search for Extraterrestrial Intelligence Is Rooted in Math

MATHEMATICAL CONCEPT: PROBABILITY

AT THIS MOMENT, an enormous group of telescopes north of San Francisco is searching the skies for signs of extraterrestrial civilizations. Named after Paul Allen, the former Microsoft executive who contributed toward its construction, the Allen Telescope Array (ATA) comprises forty-two radio telescopes that are each 6.1 meters in diameter. (There are plans for the ATA to grow to 350 telescopes.) The telescopes are being used by SETI, or Search for Extraterrestrial Intelligence, an organization based in Mountain View, California. When all the individual telescopes are in place, they will cover an area of 1 hectare, or 10,000 square meters.

Aside from the mathematics involved in engineering such a large device and in processing all the signals that the ATA will collect, math also contributed to the overall thinking behind the entire project. In 1961, Dr. Frank Drake, one of SETI's founders, created an equation that captures all the elements one would have to consider while searching for extraterrestrial civilizations capable of generating signals we can detect on Earth. Here is the Drake equation:

$$N = R^* f_p n_e f_l f_i f_c L$$

As for the terms, here are their definitions:

N = the number of civilizations in the Milky Way that emit electromagnetic radiation that human beings can detect

R^* = the rate of the formation of stars that could support intelligent life

f_p = the fraction of those stars that have planetary systems

n_e = the number of planets around each star that could support life

f_l = the fraction of those life-supporting planets that actually have life

f_i = the fraction of planets with life that have *intelligent* life

f_c = the fraction of civilizations that emit signals that we can detect

L = the length of time that those civilizations emit those signals into space

In this case, using the language of mathematics helps crystallize a group's thinking and clarifies the parameters of the project.

FERMI PARADOX

Physicist Enrico Fermi (1901–1954) was also interested in extraterrestrial civilizations and helped develop what is now known as the Fermi paradox. According to Fermi's calculations, extraterrestrials should have made contact by now. And because they haven't, Fermi famously asked: "Where is everybody?"

Do Cicadas Use Math to Protect Their Species?

MATHEMATICAL CONCEPT: PRIME NUMBERS

IS THERE ANYTHING INTERESTING to say, math-wise, about insects? If the insect is a periodical cicada, there definitely is. This bug lives in the forests of the eastern United States and belongs to the genus *Magicicada*, which includes seven species. You may have heard them during the summer as they cling to tree trunks and branches while buzzing to attract potential mates.

The mathematically relevant aspect of these cicadas is their unusual life cycle. For most of its life, the periodical cicada lives underground as a juvenile, feeding on xylem, a fluid in trees that transports nutrients. But after a certain number of years, the cicadas emerge from the soil, shed their exoskeletons and, like butterflies, transform into winged adults, ready to mate. When, exactly, do these cicadas rise out of the earth? Depending on the species, this momentous event occurs every 13 or 17 years. But these aren't just any numbers: 13 and 17 are prime numbers, divisible only by themselves and 1; other examples of prime numbers are 5 and 11 (see entry #86).

Why do cicadas incorporate primes into their life cycles? Some scientists conjecture that cicadas developed their prime-based lives to outsmart predators. To see how the primes would defend cicadas against other animals, imagine that the cicadas instead emerged from the ground every six years. Because six is divisible by 1, 2, 3, and 6, any animal whose life cycle incorporated those numbers could expect to sync up with the cicadas' life cycles, leaving each new cicada generation vulnerable to attack. But since the cicadas emerge on a schedule based on prime numbers, the odds that another animal's young would be looking for food just as the new brood of cicadas was maturing would be reduced. Periodical cicadas, in essence, might be using prime numbers as a defense mechanism.

BAMBOO

Cicadas are not the only organisms that have reproduction schedules that might be tied to predators. The Japanese bamboo, for instance, only flowers every 120 years. Scientists conjecture that the long lag time evolved so the rodents that feed on the seeds would die out between flowerings, in effect regulating the rodent populations. And it doesn't matter where they are planted: they flower in 120 years, like clockwork.

100

Base 2

MATHEMATICAL CONCEPT: NUMBER SYSTEMS

THE NUMBERS WE USE every day are so familiar that we sometimes don't appreciate certain aspects of them. But if you were to look closely at our numbers, you would begin to notice peculiarities. For instance, our numbers are made up of only ten numerals: 0, 1, 2, 3, 4, 5, 6, 7, 8, 9. But why is there no special numeral indicating ten? Or eighty-five? Or three thousand one?

The answer is that our numbers use a system called Base 10. In this system, each position in a number stands for a particular power of 10. If there is a 1, 2, 3, or any other numeral in that position, we multiply that number by that power of 10. Let's look at an example:

$$642$$

In this number, the 2 is in the first spot on the right. This spot is the ones, or 10^0, position. So, we multiply 2 by 1. The next number to the left is 4. The 4 is in the tens, or 10^1, position, so we multiply 4 by 10. And the 6 is in the hundreds, or 10^2, spot, so we multiply 6 by 100. So, the numeral 642 stands for (6 times 100) plus (4 times 10) plus (2 times 1), or six hundred forty-two.

Is there another number system that uses a base that isn't 10? Yes, and in fact it undergirds every computer in existence! It's the Base 2 system, and just like the Base 10 system, it relies on positions. The furthest position to the right is the 2^0, or ones, position. The next position to the left is the 2^1, or twos, position. To the left of that are the fours, eights, and sixteens positions. But unlike the Base 10 system, the Base 2 system only has two numerals: 1 and 0. So, if we wanted to write five in Base 2, we would write this:

$$101$$

There is a 1 in the fours position, a 0 in the twos position, and a 1 in the ones position; together, they equal five. (Computers use Base 2 primarily because it's easier to distinguish between two states—1 and 0, corresponding to on and off—than among ten of them.)

ONES AND ZEROES

Ones and zeroes can also be used to encode letters. Binary code translates each letter into a unique string of eight ones and zeroes. Variations of this kind of code were explored by sixteenth-century thinker Francis Bacon, and by mathematician and philosopher Gottfried Leibniz in the seventeenth century.

INDEX

Index

Index

Index

ABOUT THE AUTHOR

RAPHAEL ROSEN caught the science-writing bug while working at the Exploratorium—San Francisco's hands-on museum of science, art, and human perception. There he was inspired by the way the Exploratorium's exhibits communicated science ideas clearly and in a down-to-earth fashion. (The giant ball bearing and wave interference exhibits especially stand out in his memory.) Always interested in the ways in which science and writing intersect, he has been inspired by Jerry P. King, author of *The Art of Mathematics*, as well as K.C. Cole, Philip Hoare, and Bryan Magee.

He has written for the NASA Spitzer Space Telescope mission, as well as for *www.space.com*, where he explored the aesthetics of space telescope images and covered recent research news in the space community. He wrote about arts and cultural events for the *Wall Street Journal*, and interviewed paleontologist James Horner for a piece about fossil auctions that ran in *EARTH* magazine. Rosen also wrote a story about Maxfield Parrish's lavishly illustrated undergraduate chemistry notebook for *SciArt in America*, and contributed short news items to *Scholastic Science World*. He has written for *Discover* and *Scientific American*, and authored a children's book about outer space.

Rosen has a master's in specialized journalism from the University of Southern California and a bachelor's in philosophy from Williams College. Originally from Winston-Salem, NC, where he frequently visited the Nature Science Center, he currently lives in New York City and can be found online at *www.raphaelrosen.com*.